Sticking Together
The Science of Adhesion

CW00674594

Sticking Together
The Science of Adhesion

Steven Abbott
University of Leeds, UK
Steven Abbott TCNF Ltd
Email: steven@stevenabbott.co.uk

ROYAL SOCIETY
OF **CHEMISTRY**

Print ISBN: 978-1-78801-804-3
EPUB ISBN: 978-1-83916-015-8

A catalogue record for this book is available from the British Library

© Steven Abbott 2020

All rights reserved

Apart from fair dealing for the purposes of research for non-commercial purposes or for private study, criticism or review, as permitted under the Copyright, Designs and Patents Act 1988 and the Copyright and Related Rights Regulations 2003, this publication may not be reproduced, stored or transmitted, in any form or by any means, without the prior permission in writing of The Royal Society of Chemistry or the copyright owner, or in the case of reproduction in accordance with the terms of licences issued by the Copyright Licensing Agency in the UK, or in accordance with the terms of the licences issued by the appropriate Reproduction Rights Organization outside the UK. Enquiries concerning reproduction outside the terms stated here should be sent to The Royal Society of Chemistry at the address printed on this page.

Whilst this material has been produced with all due care, The Royal Society of Chemistry cannot be held responsible or liable for its accuracy and completeness, nor for any consequences arising from any errors or the use of the information contained in this publication. The publication of advertisements does not constitute any endorsement by The Royal Society of Chemistry or Authors of any products advertised. The views and opinions advanced by contributors do not necessarily reflect those of The Royal Society of Chemistry which shall not be liable for any resulting loss or damage arising as a result of reliance upon this material.

The Royal Society of Chemistry is a charity, registered in England and Wales, Number 207890, and a company incorporated in England by Royal Charter (Registered No. RC000524), registered office: Burlington House, Piccadilly, London W1J 0BA, UK, Telephone: +44 (0) 20 7437 8656.

Visit our website at www.rsc.org/books

Printed in the United Kingdom by CPI Group (UK) Ltd, Croydon, CR0 4YY, UK

Preface

When I accidentally knocked over a nondescript jug and broke its handle, I assumed I could throw it in the bin. But no. It was my wife's orchid-watering jug and she made it clear that she would be most unimpressed if the book I was writing couldn't tell me the best way to fix it. That incident caused me to re-think my priorities. Of course, I was trying to make my writing interesting and of course I was trying to make the science clear and correct. However, was the book also going to be useful? It turned out in this case that what I'd already written about fixing handles was helpful and I could make a rational choice of a low viscosity (cheap) superglue, using a puff of breath to ensure enough catalyst (water) and a quick, decisive press and hold. I was also alert to the downsides of super-glue – there was no going back. What I failed to do was practice the swift motions required for gluing the second piece of the handle. The result is a solid handle, fit for watering orchids, but with a crack that is more unsightly than it should have been.

I also wanted the book to be more than just a book. A useful working assumption is "If it's not on YouTube then it doesn't exist". To which we need to add, "If 'it' is more than 5 minutes then no one will watch it so it still doesn't exist". So, the book comes with its own YouTube playlist (rsc.li/books-sticking-together) of informal, short videos which bring some of the

Sticking Together: The Science of Adhesion
By Steven Abbott
© Steven Abbott 2020
Published by the Royal Society of Chemistry, www.rsc.org

science to life. You'll find a video about what's being discussed wherever you see the icon:

I already had an extra resource for the expert reader. Many times in the book, I provide a link to an app page on my large Practical Science website. The page does not just contain the science and formulae; it brings the science to life with sliders, graphs and outputs to allow the user to get a feel for what the science means. Modern web infrastructure is so powerful that even on your smartphone you can explore some fairly complex science via these apps if you wish to know a little more.

Going back to the question of whether the book would be useful, I needed to learn from those with experience of using adhesives in many ways unfamiliar to me. I am especially grateful to Rob Johnson, Denis MacSweeny, William Newman-Sanders, Jonathan Parker and Lt-Col Chris Urdzik of the USAF for patiently guiding me through their extensive experiences with adhesives in the real world.

I needed access to some professional testing equipment for the YouTube videos. I am most grateful to Elliot Allen, Vivian Christogianni and their colleagues at Sugru for making their lab and equipment available.

Dr Steven Howe provided the idea, and the hands-on skills, to demonstrate the complexities of the lap shear test.

I wanted to avoid "sciencey" illustrations and was delighted that Mark Abbott was able to translate my technical illustrations and scribbled suggestions into his distinctive informal-looking illustrations that make up the majority of graphics in the book.

I needed a style guide who had a rare combination of a good grasp of English style along with the ability to provide an informed critique of the scientific story as it unfolded. I'm grateful that Sebastian Abbott took on that tough dual role and provided innumerable improvements to the style and content.

Ella Abbott was an able assistant for many of the videos and played an important on-screen role in one of them.

Drew Gwilliams and Katie Morrey at the Royal Society of Chemistry were most helpful and supportive in steering the process from its earliest tentative stage through to completion.

You will encounter me throughout the book. I've written it in a first-person style because I hate hiding behind a supposedly objective neutral authorship. I've worked for most of my life in real jobs in industry while having the privilege of being a Visiting Professor at the University of Leeds which has given me access to the world of academia. Since 2009, I have been an independent scientist working on topics that interest me, learning new things, helping solve technical problems around the world and, as much as possible, and giving away my knowledge via free apps and eBooks. You will find that a lot of the bad science that is out there irritates me. My chief irritation is that the bad science doesn't help anyone solve problems. The bad science is unnecessary because the real science of adhesion is not at all difficult and has the distinct advantage that it makes you much better at sticking things together.

For those who want to read about the science in more detail, with extensive links to the apps to make it easier to understand, my book *Adhesion Science: Principles and Practice* is available from DesTECH publications.

Contents

Sticking Together: The Science of Adhesion
By Steven Abbott
© Steven Abbott 2020
Published by the Royal Society of Chemistry, www.rsc.org

Glossary

Accelerator	An additive to make the adhesive set faster
Adherend	The thing being stuck via the adhesive
Adhesion promoter	A low-level additive that increases the adhesion
Alcohol	A molecule containing an –OH group
Amine	A group based on nitrogen and two hydrogens, shown as $-NH_2$
APTES	A common adhesion promoter that is both an amine and a silane
Backing tape	The carrier film for a PSA. Sometimes just called "backing"
Butt test	Pulling apart in a straight line
Catalyst	Something which makes a reaction go faster without itself being consumed by the reaction
Coalescence	Small, semi-solid particles flowing together to give a continuous solid
Crack energy	The energy trying to open a crack across the interface
Creep	Slow movement in a joint under a constant load
Curing	Setting solid via polymerization
Dahlquist	PSAs must be soft enough to meet the Dahlquist criterion for quick tack

Sticking Together: The Science of Adhesion
By Steven Abbott
© Steven Abbott 2020
Published by the Royal Society of Chemistry, www.rsc.org

Dispersant	A molecule added to a formulation to keep particles separated and free flowing
Dissipation	Converting work into useless heat – this soaks up crack energy and increases adhesion
DLVO	A theory of why small particles will or will not stick to each other, named after Derjaguin & Landau and Verwey & Overbeek
G' & G''	Measures of the elastic (G') and fluid (G'') resistance to shear forces
Griffith's Law	The tendency of a crack to form increases with the size of defects in the system
Hydrophilic	Likes to be with water, dislikes oil
Hydrophobic	Dislikes water, likes to be with oil
Hydroxyl	A group made from one oxygen and one hydrogen, shown as –OH
Interface	The line between two adhesive components, e.g. between adherend and adhesive
Ion	A charged molecule or atom. Anions (negative ions) can polymerize cyanoacrylates
Molecular weight	For polymers this indicates how many monomer units were assembled and, therefore, the length of the chain
Monomer	A molecule that can react with other monomers to form a polymer
Peel test	Pulling up from the joint
Polymers:	
EVA	(Poly)Ethylene vinyl acetate
EVOH	(Poly)Ethylene vinyl alcohol
PE	Polyethylene, commonly called polythene.
HDPE/LDPE	High Density PE and Low Density PE
PP	Polypropylene
PMMA	Polymethylmethacrylate, commonly called Perspex or Plexiglas
PC	Polycarbonate
PET	Polyethyleneterephthalate, commonly called polyester
PDMS	Polydimethylsiloxane, commonly called silicone

PTFE	Polytetrafluoroethylene, commonly called Teflon
PVOH	Polyvinylalcohol
PVA	Polyvinylacetate. Confusingly PVA can also mean Polyvinylalcohol
PVB	Polyvinylbutyral
PVP	Polyvinylpyrrolidone
PSA	Pressure Sensitive Adhesive – common adhesive tape
Radical	A reactive molecule that produces a new radical after reacting with a monomer
Retarder	An additive to slow down the setting speed of the adhesive
Shear test	Pulling apart across the joint
Stefan's law	Squeezing a blob of adhesive gets much harder as its thickness decreases
Stress	Force acting across an area
Strain	Amount of stretch caused by a stress
Surfactant	"Surface active agent" used in cleaning ("detergent") and creating emulsions ("emulsifier"). Has a hydrophilic head and hydrophobic tail
Tack	That easy to know but hard to define feeling of instant adhesiveness
-TES & -TMS	Triethoxy and Trimethoxy silanes, very useful adhesion promotion groups
TiO_2	Titanium dioxide – a key whitening pigment
van der Waals	The force that attracts all molecules and surfaces to each other
vdW	Abbreviation for van der Waals
VOC	Volatile Organic Compounds – typically solvents other than water
WLF	Williams, Landel and Ferry theory describing the equivalence of temperature and time

CHAPTER 1

Introduction

Modern society would fall apart without adhesives.

Your smartphone has at least 25 adhesive joints in it, with the all-important glass screen and the display held with adhesives, as are the chips, microphones, speakers and aerials inside. The screen protector is also stuck on.

Now step into your car. Manufacturers are under intense pressure to reduce weight. Every screw or rivet is a potentially avoidable weight (yes, automobile manufacturers agonize over each gram in each component) and the interior of a modern car is held together by clips (for removable parts) and adhesives. External parts such as bumpers (fenders) used to be metal. Now they are made of polypropylene, strongly adhered to the metal shell. Even the metal shell itself is rapidly heading for history as plastic/carbon fibre takes over, held together with adhesives.

What about something really important such as an aircraft? The interiors are stuck together – the weight, inconvenience and aesthetics of screws and rivets make them unacceptable. The rivets we see on the outside give the impression that they are the key structural element, yet they are now mostly for backup or are redundant. Even the (by now) old all-metal aircraft are largely held together with adhesives, while the latest generation of civil aircraft such as the lightweight, efficient Boeing Dreamliner and Airbus 350 are predominantly glued together.

Sticking Together: The Science of Adhesion
By Steven Abbott
© Steven Abbott 2020
Published by the Royal Society of Chemistry, www.rsc.org

And yet, the hidden nature of adhesives (if you can see them, they probably haven't been applied properly), allied with their avoidability (people often prefer to try other methods first), has led to a rather dismissive attitude to the whole technology. When I mention my interest in adhesives to friends and neighbours, the general response is that adhesives are trivial, unsatisfactory or excessive, i.e. they are only an occasional feature of their lives, they often don't work well and can be (like my wife's orchid jug mentioned in the Preface) unsightly when used; as such they are not of any great importance. There is a medieval notion that to "really" put things together you need nails, screws, rivets and welding.

Talk to anyone who really needs to put things together and you find a horror of nails, screws, rivets and, with the exception of steel for which it is wonderful, welding. A good way to make anything fail is to put a hole in it or to heat parts of it while other parts remain cool. When things are held together by nails, screws or rivets, stresses are focussed on the hole and crack cans start there. Those fixing devices become instant sources of potential failure. The bio-inspired hook-and-loop type of fasteners developed and commercialized under the name Velcro are an ingenious variant on the old hook-and-eye system. Although they are a welcome addition to the world of fasteners thanks to their ability to distribute a load over a wide area, because they are purely mechanical, they feature no further in this book. And, of course, these fasteners themselves are often attached via a strong adhesive.

Adhesives, in contrast to conventional fixings, spread the load and do not require you to add defects to your object. This means that they are used just about everywhere.

The high-tech adhesives industry is very much alive and well. For phones, cars, aircraft, medical devices, construction, packaging – just about everywhere – customers demand not so much better adhesives as better adhesive *systems* that deliver a package of benefits.

As we will discuss later, there is a constant set of trade-offs to be battled. Prioritizing strength and hardness can lead to systems that are too brittle and cannot cope with shocks. Making things flexible and resilient can lead to compromises in resistance to those steady, long-term forces that can produce creep (slow deformation) within the bond.

Providing controlled low adhesion is especially tricky for consumers.

- People like easy-open packages and get frustrated if they are too hard to pull or give a raspy "stick-slip" release. But no one likes it when they are so easy to open that this has happened during storage.
- Similarly, we all want hooks that are easy to stick to a surface and that securely support a large weight. But we also want them easily removed and repositioned with zero damage to the surface.

In addition to the functional challenges, everyone says that they want "green" adhesives. Going back to the bad old days of collagens and milk protein, which will be discussed in Chapter 1, is not a sensible option, both technologically and in terms of those who would object to animal products in, say, their smartphone. Trying to replace petroleum-based adhesive ingredients with those made from renewable resources is neither straightforward nor provably green. A life cycle analysis that takes into account, for example, the water and fertilizer used in growing a crop, the energy needed to harvest it and transport it, plus all the processing and waste disposal does not automatically show that such materials are preferable. And that is before taking into account the alternative use of that land: growing crops for food. This is not just my personal view; some major EU studies on, for example, bio-based plastics confirm that there are often major downsides to switching from petrochemical alternatives. I have had conversations with urethane manufacturers who are frustrated by the fact that the variability of bio-based raw materials has a serious impact on performance, with whole batches sent to waste (definitely not green) because the adhesive performance was sub-standard.

In the foreseeable future the greenness of the adhesives will consist of doing more with less; minimizing the amount of adhesive in each bond, which in turn requires greater precision in the parts to be adhered. You can't use a 1 μm layer of adhesive if the parts have a 5 μm roughness. And as more and more parts of products such as cars and aircraft are made from (fibre-reinforced) plastics, the environmental upsides from using a small amount of a high-performance "non-green" adhesive far outweigh any environmental downsides of the adhesive itself.

Nature has many examples of ingenious modes of adhesion that excite a lot of well-intentioned publicity when scientists find ways to use similar principles. As we shall see in the final chapter, the principles are usually more to be admired than to be copied. If I am flying in an aircraft mostly held together by adhesive, I far prefer a wholly synthetic, high-tech, apply-and-forget adhesive to a "smart" bio-inspired one that needs a constant supply of chemicals in water to keep it in good shape.

1.1 WHICH ADHESIVE SHOULD I USE?

What everyone, manufacturer or user, wants to know is whether this specific adhesive will do a great job on this specific adhesive problem. We have the whole of Chapter 3 to see how adhesion is tested in the lab. What about testing it at home?

Mostly we have one-off jobs such as sticking a drawer handle back on, or fixing the leg of a chair. Our "test" therefore is whether the job worked out OK.

If we *regularly* do an adhesion task then we can try out a few different adhesives and a few different application methods, working out which is the best balance of cost, speed and effectiveness. Most of the time we don't have that luxury – we have to make a one-off decision to use *this* adhesive, applied in *this* manner to stick *these* things together. If it works, no one will ever comment. If it fails, then you risk anything from embarrassment to significant loss.

One aim of this book is to help you make the right choice of adhesive for any given job, and in Chapter 6 we review many common systems for sticking A to B. To make the right choice you need a key fact that is missing from most accounts of adhesion and adhesives. Here it is:

"Adhesion is a Property of the System".

You will have no problem remembering this phrase because, I make no apology, it appears many times throughout the book.

The biggest mistake any of us can make when thinking about an adhesion problem is to focus on the adhesive, rather than the system. If your system is going to involve lots of peel, then, as we shall see, don't bother with a superglue. If your system involves lots of shear, then (all this will be explained clearly later) the strength of the bond depends as much on the thickness and modulus of the adherends (the things you are adhering) as on the adhesive. If you can increase the thickness and/or modulus of the adherends you are already improving things, even without thinking about the adhesive.

Then you need to worry about speed, and its equivalent, temperature (yes, the two are strongly inter-related as we shall discover). If your problem is long, slow loads and/or higher temperatures then a "strong" adhesive will be right. If the problem involves short, sharp shocks (and/or cold temperatures) then a strong adhesive might be catastrophically brittle and you will need something far more forgiving. The common PVA wood glues are used extensively not because they are amazingly strong (which they're not) but because they are amazingly forgiving when the woods in the joints (they might be different types of wood or in different grain directions) expand or shrink with the rise and fall of humidity.

Adhesives have, as we will see, moduli, viscosities, glass transition temperatures, curing speeds, degrees of cure, cross-link densities. Each of these can be measured and a supplier could, if necessary, give you all those values. What no supplier can give you is a meaningful statement about how strong it is, because no adhesive has "a" strength, because Adhesion is a Property of the System.

A supplier *can* say that this adhesive can survive $X \, N \, m^{-2}$ when tested against Test Standard XY92, and you *can* compare that to a different adhesive tested against the same standard and it is *possible* that the test is relevant to the type of loads you are trying to resist. But we are not likely to have such a situation in our day-to-day fixing jobs.

I now want to flip all these negatives into a positive. You are the one who knows what you are sticking to what, for what reasons, and you know the sorts of assaults the joint will receive over its required lifetime. You also know the restrictions of contamination, access space, time, temperature, weather, sunlight. You are the world expert on your system. Now that you know that Adhesion is a Property of the System, and that you are the expert on that system, you don't have to be taken in by adverts for glues that work only under the precise conditions created for the advert. You don't have to be fooled by statements like "sticks anything to anything", with a little asterisk pointing you to a set of disclaimers in small print.

With help from the chapters that follow, you will be able to:

- Understand what will or won't help with surface preparation
- Look at advertising claims with a healthy scepticism
- Choose between a "strong" or a "tough" adhesive
- Choose between a thick or thin layer of adhesive
- Choose between a good general-purpose adhesive and one (allegedly) specifically designed for your sort of system
- Know whether the adhesion promoters present in some adhesives will help (and how) or hinder (and why)
- Understand why too much of a good thing is a bad thing
- Find out how to *reduce* adhesion when you need to

You cannot do these things well if you assume that everything is down to the adhesive. By understanding the system, you go a long way towards understanding how to get the best out of what you have to hand.

1.2   HOW THE BOOK UNFOLDS

We start by admiring those pioneers of adhesion who managed to take crude raw materials such as birch bark tar or boiled bones to create really rather impressive adhesives. We then switch to some necessary basics to become familiar with the few core ideas needed to understand the rest of the book. By looking at how geckos manage to stick to walls, we see the sort of adhesion we mostly *don't* want, getting ready to find out how to get the (usually) strong adhesion we *do* want. But before getting to strong adhesion we need to know how to measure if our adhesion is strong. Because adhesion is a property of the system, this is by no means obvious. Then we can get to understand how strong adhesives work (and when they will fail). Because much of strong adhesion depends on *strong* polymers, we need then to switch to pressure sensitive adhesives (common tapes) that give strong adhesion thanks to very *weak* polymers. What unites the strengths of both types of adhesives is that they each manage to dissipate the energy of a potential crack; adhesion is much more about dissipation than it is about "strength". That completes the next five chapters and provides all the principles we need. The final five chapters are about specific systems and how they work with the principles we've worked hard to understand.

CHAPTER 2

Background Ideas

We take adhesion for granted because most of the time it just works. We only notice it when it goes wrong – when the thing we fixed at home breaks again or when the removable adhesive isn't so easy to remove. One reason for us taking adhesion for granted is that modern adhesives are so good. It wasn't always like that.

Take yourself back hundreds of thousands of years. Adhesion is now a matter of life or death. If you can reliably stick a flint arrowhead into a slot at the front of an arrow (and stick some feathers onto the back) then you will be able to eat tonight (Figure 2.1) – if not you starve. How hard can it be? Just get some sticky stuff and, well, stick it. Ah, that pine tree has some sticky stuff, let's try it. The real sticky stuff is too soft, so you try the hardened version, using a fire to melt it. Pour it around the arrowhead and stick and, when it has cooled take a test shot. As soon as the head touches the target, the brittle pine glue shatters. You won't eat tonight. After a few decades or centuries, chance or a course in advanced nanotechnology leads someone to mix the pine resin with the right sort of finely-ground charcoal. Now the adhesive is shatter-proof.

This story of the development of a working adhesive captures the frustrations of adhesive developers today. It is easy to make an adhesive that is too soft, it is easy to make one that is too hard and brittle. Finding the right balance remains a deep challenge,

Sticking Together: The Science of Adhesion
By Steven Abbott
© Steven Abbott 2020
Published by the Royal Society of Chemistry, www.rsc.org

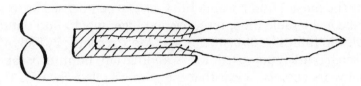

Figure 2.1   Gluing a flint arrowhead into a wooden haft is a big technical challenge. It is impressive that Neanderthals managed to do it with birch bark tar that was neither too soft nor too brittle.

especially because, as emphasized in this book, adhesion is a property of the system, not just of the adhesive.

There has never been a shortage of things that might produce a useful glue. Anyone who has overcooked any starchy food has created an adhesive. You don't even have to cook it – a paste of wheat or corn starch creates an equivalent glue. It just isn't very good, especially if the joint ever gets wet. Bacteria and moulds love to feast off the nutrients, so a starch-based glue will go off in storage and a joint might fail via mould growth.

Birch bark tar (pitch) has been a wonderful adhesive for thousands of years and birch forests are common. To get the tar you have to heat the bark; the problem is that if you heat it in the presence of oxygen it gets burned to something useless. If you were in a birch forest and had to do anaerobic ("without oxygen") heating of tons of birch bark to make large quantities of adhesive for your tribe, how would you go about doing it if you did not have access to modern tin cans?

Birch bark tar is known to have been used by the Neanderthals; archaeologists have studied tar-hafted arrow heads and discovered nearby lumps of tar ready to be used for gluing. One team of scientists, therefore, had to have a go at working out how the Neanderthals might have done it. The team tried out various methods, using embers, holes in the ground or a more complex raised structure, seeing how much usable tar they could create for a given amount of time, bark and firewood. They looked at wrapping some birch bark in fresh fibres and surrounding the bundles with embers; they dug a hole, placed a birch basket at the bottom, laid on some bark then threw in some hot embers. And the team tried something similar but covering it all with earth and lighting a large fire. In all three cases it turns out that you (can) get respectable quantities of tar, with the covered structure giving the most. I find it wonderful that 21st century scientists will spend considerable time, resource and ingenuity to find out how people did things ~100 000 years ago. More recently, by looking at the birch bark gum used for Scandinavian hunting weapons 10 000 years ago, it is clear that the gum was first chewed. Teeth marks show that both children and adults were involved in the process, and female DNA extracted from the gum shows that this was not a male-only activity.

Then there are the glues from dead animals: fish, rabbits, horses, cattle. Boil up the skins, cartilage, hooves and bones and. . . you can make gelatine, which is not useful, except in food. The trick is to control the boiling so that the collagen, the tough protein in all those parts of the animal, is broken down sufficiently to become a meltable glue yet not so much as to be reduced to gelatine (Figure 2.2).

It is a genuinely difficult challenge to get these processes to work at all – imagine having to provide the quality control to ensure that they work day after day!

One useful additive to cartilage-based glues sounds vaguely amusing to us now: urine. It is at first surprising how often urine appears in ancient recipes and processes. We now know that the main chemical in urine, urea, is one of the few molecules that interacts strongly with proteins such as collagen to make them more soluble. Camel urine was for centuries a hair-care essential because it allowed the keratin protein in hair to be made flexible before being shaped into whatever was the current fashion. Urea is frequently used in skin cream formulations and is an essential part of "natural moisturizing factor" created by our skins. As

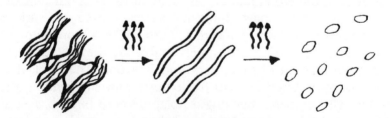

Figure 2.2   The tough, insoluble collagen from bones, skin *etc.*, (left) needs to be boiled sufficiently to break it down into soluble glue molecules (middle) without being broken further into smaller lumps that constitute gelatine (right).

with all complex products, adding too little or too much urea would cause problems; it would be interesting to know what quality control procedures were used to ensure consistent urea additions from such a variable raw material.

How do we know what sorts of glues were being used a long time ago? Archaeology shows that locals used the resources to hand, and some of them are easy to identify even after millennia: Aztecs used rubber; Mesopotamians in 4000 BC used bitumen to attach ivory eyeballs to statues; and from the Neanderthals onwards, the plentiful supply of birch bark in Northern forests drove the use of pitch. In South Africa, Yellowwood pitch did the same job 60 000 years ago. Collagen has a distinctive protein "signature" – its mix of peptides is very different from the majority of proteins. If you come across some decorated Chinese burial staffs from 3500 years ago and have access to a modern pro-teomics lab it is not too hard to find some bits of hardened adhesive which, despite some degradation over time, show the distinctive collagen signature. Carvings from ~1450 BC in Thebes in Egypt show glue pots being used for laminating veneers; analysis of Egyptian pot fragments suggests, again, collagen glues.

Those artisans who made furniture fit for Egyptian pharaohs needed adhesives with the right consistency to keep the chairs and tables from falling apart during the hot, dry periods when the wood and glue dried out, and during the hot, humid periods when the wood was swollen and mould and bacteria would feast on the peptides and proteins in the glue.

Similarly, for the Greek and Roman nobility, an absolute "must have" was furniture that combined ornate wooden sur-faces with at least some degree of practicality. With the right collagen glues, it became possible to apply thin sheets of intri-cately patterned veneer to an otherwise dull table. To be even fancier, different coloured woods could be stuck together in patterns, the art of marquetry.

The joints and veneers those Egyptian, Greek or Roman artisans worked on allowed them to easily spot any problems caused by the weather and, with their collagen-based adhesive systems, to make timely interventions. This is because their glues had a key advan-tage missing from our current high-tech versions: with modest amounts of heat and water or steam the artisans could get a joint to fail in a controlled manner so they could replace or repair it, returning the furniture to its former glory. To this day, violin makers use animal glues so that it is easy to take the instrument apart and fix any problems. Those famous Stradivarius violins will have been taken apart many times over the centuries while still retaining their glorious sounds. If someone used a modern adhesive, taking the violin apart to repair it would probably destroy it.

In societies where milk was common, it was rather easy to separate out a protein, casein (the word is related to "cheese"), by allowing the skimmed milk (all the cream separated) to go sour or by deliberately adding an acid. Casein, like collagen, is a protein and on its own forms a solid film with some adhesion, but that's not how you make a good glue. In fact, the process is quite tricky. The casein has to be washed, dried and ground into a powder. Then it has to be stirred up – and here's the thing – if it is stirred up with lime it makes a great, water-resistant adhesive, as long as you can act fast enough before it sets solid; if it is stirred up with sodium hydroxide it has a long pot life and gives a strong bond but not a water-resistant one. The right mix of lime and sodium hydroxide gives you a good working life and adequate water resistance. That doesn't sound so hard, and with internet access to recipes and easy purchase of pure lime and pure sodium hydroxide it really is not so hard. Now transport yourself back in time where it wasn't even clear what sodium hydroxide actually was. And no one knew that lime contained calcium ions that love to bind strongly to carboxyl groups on the protein (Figure 2.3). These simple, natural glues are neither

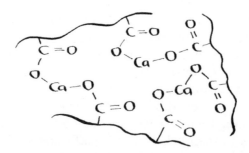

Figure 2.3   The carboxylic acid groups, CO_2^-, on the casein interact with the calcium ions, Ca^{++}, from the lime to provide strength and water resistance.

simple nor natural. Who first came up with the idea of mixing casein with lime and sodium hydroxide – and was able to successfully reproduce its excellent performance?!

A different, and abundant source of protein suitable for glue came from the blood from slaughterhouses. Haskell, a Michigan businessman, had access to large volumes of blood from the Chicago stockyards and plenty of cheap timber to stick together with the "blood glue" to create Haskelite (what we would now call plywood) for vehicles, canoes and aircraft.

There is one more protein adhesive with which I have had personal experience. Applying gold leaf is an ancient craft, and egg white albumin (a protein) is a key ingredient of the adhesive. I needed to decorate a harpsichord and was taught by an expert how to apply the leaf. I followed all the tricks that had been handed down over many generations; for example, huffing onto the albumin adhesive just before applying the leaf to make the adhesive slightly more tacky. I suspected, however, that many of these traditional steps must be unnecessary, and I worked out my much smarter way to do it. When I tried out this "smarter" way, the results were a disaster, so I swiftly reverted to the proven methodology.

Adhesion is not just about sticking two or more surfaces together; it can also be about protecting a single, specific surface, in the form of a sealant or a varnish. An intelligent mind like Leonardo da Vinci's would not have been happy to see one of his paintings leave his studio without a hard, clear protective coating to help it survive life in a draughty palace or cathedral. His paints were oil paints, dispersed in solvents such as turpentine. He needed a tough, compatible polymer coating that could integrate itself into the paint as it solidified. There are such polymers available, in the form of amber, shellac, gum Benjamin and others, that are soluble in such solvents and thus provide the right compatibility. The problem (and it is the same today) is that any solvent good enough to unite the two might eat too much into the paint layer and destroy it. The artist had to find blends of solvents that had the right "bite" into the painting – not too much (destruction) or too little (the varnish falls off over time).

2.1 WHY BOTHER WITH ADHESIVES?

Civilizations can get by without large-scale adhesive use. We know this because although the Egyptians, Romans, Greeks and Chinese had large-scale adhesive industries for, say, furniture making, medieval Europe coped OK for centuries without them, having lost many of the technologies and not having an urgent need to re-create them. You can make timber structures with holes and pegs, you can melt, hammer and weld metal components together, you can sew, lace, hook and tie clothing together. For decorations you can probably make some small amounts of sticky stuff that do the job, without the need for a significant industry.

It wasn't until the 16th century that European princes wanted fancy cabinet making and laminated woodwork, and musicians wanted large, delicate instruments. The demand from the elite ensured that the art and science of adhesives was redeveloped. The first large glue factory (with horses as a key raw material) was founded in Holland in the late 17th century. In the 18th and 19th centuries patents for fish and casein glues were published.

A kind researcher at the British library tracked down for me the earliest known British patent for a glue. The inventor, Peter Zomer, was from the Netherlands, and the patent is really about getting both the "train oil" (whale oil; the drops are seen as being like tears, which are *traane* in Dutch) and a fish glue from the Greenland whaling industry:

> British Patent #691. *Whereas His most Sacred Majesty George the Second ... bearing the date at Westminster, the Twenty-third of May [1754] I Peter Zomer by petition humbly represent to His Majesty that I had found out and invented "A Method of Extracting and Making from the Tails and Finns of Whales, and from such Sediment Trash and Undissolved Pieces of the Fish as were usually thrown away as useless and of little or no Value by the Makers of Train Oil, after the Boiling of the Blubber of such Fish, a Sort of Black Train Oil, and afterwards of Making from the Remains of such Tails, Finns, Sediments, & Undissolved Pieces a Kind of Glue called Fish Glue."*

Not long after, the relatively sophisticated nature of the French adhesives industry was described by M. Duhamel du Monceau in

The Art of Making Various Kinds of Glues, 1771. This fascinating book was translated into English in 1905 by the J. Paul Getty Museum and is easily found on the internet. Of special interest is that a particularly high-class fish glue (as opposed to the rather poor stuff made by Zomer) was available only from Russia and M. du Monceau took the trouble to find out what it was. In modern language, it was the swim bladders – rather, pure collagen – of beluga sturgeon.

There is also an example of how a good source of glue (collagen) became a poor one through market forces. Those who fancy any form of "good old days" using "natural" adhesives produced by happy artisans may not enjoy this little snippet from the book:

> *"The feet of oxen, formerly esteemed, are now looked on as one of the bad materials that can be employed, & especially since the Butchers have begun to carefully remove a tendonous part of them, called the small nerve, or the shin nerve, that they sell by weight, & rather dearly for the production of a kind of oakum which is useful for caulking the panels of carriages, or to make suspension straps for carriages. When the feet are thus stripped of this tendonous part, they produce only a mucilaginous substance which is not suitable for making good glue; & if anyone makes use of them, it is because of their low price."*

As M. du Monceau's book indicates, by the early 19th century, we enter the world of industrial adhesives that take up the rest of the book.

2.2 TECHNICAL BASICS

Before we start, we need to make sure we have the right basic ideas and language to understand what makes a good, or poor, adhesive system. Be assured that what follows is rather gentle, yet at the end you will be able to understand what's really going on when we stick things together. We take it step by step, with none of the steps being especially difficult.

The trickiest, final, section covers the science of the polymers used as adhesives. The emphasis is on the few important principles (e.g. what polymerization is, what a crosslink is) that anyone can grasp. Readers who dread chemistry need not worry.

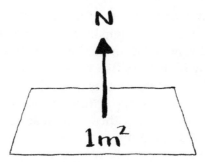

Figure 2.4   Force, in N, is applied over an area in m^2. 1 N m^{-2} is called a Pascal, Pa.

2.3   UNITS

We have to agree on a few measurement units and technical terms. Most readers will be familiar with them as they aren't too exotic.

For *length* we will use metres, m, millimetres, mm, micrometres, μm and nanometres, nm, each 1000× smaller than the previous. The unit of centimetres, cm, doesn't fit into that nice scheme but is so common that it has to be included.

For *time* we will go down to μs, ns and ps, micro, nano and picoseconds, one millionth, billionth and thousand billionth of a second.

For *weights* and *loads* we will use kilograms, kg, grams, g and Newtons, N, which, if a weight is involved is just weight times gravity. For our purposes, gravity is 10 m s^{-1} s^{-1} so 1 kg is 10 N.

Force per unit area has the units of N m^{-2} which is also expressed as Pascals, Pa (Figure 2.4). One Pa is rather small, so we often have kPa, MPa and GPa for kilo, mega and giga, 1000, 1 million, 1 billion Pascals.

If you pull, say, a piece of plastic which has a cross-sectional area of A with a force *F*, then you will get a fractional increase of length (the change in length divided by the original length), ε (Figure 2.5). The force per unit area is the *stress* in Pa. The fractional increase in length, *strain*, has no units. If you divide stress by strain you get *modulus* which gives an idea of the strength of the material. Modulus is also measured in Pa, as strain is dimensionless. Dividing by a small number gives a larger number, so the smaller ε for a given stress, the larger the

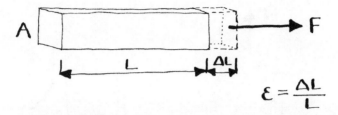

$$\varepsilon = \frac{\Delta L}{L}$$

Figure 2.5   When we apply a Stress, Force/Area (*F/A*) in Pa we get an elongation, ΔL of the original length L. Their ratio is ε, which is the Strain, which is unitless. The ability to resist stresses is the Modulus, Stress/Strain, also in Pa.

modulus, which makes sense because a stronger material will stretch less. The modulus of typical polymers lies in the 1–4 GPa range, steel is 200 GPa. We will find later in the book that adhesion doesn't necessarily rely on brute strength; if the adhesive on a strong household tape has a modulus *greater* than 0.3 MPa, it does not work (it has no stickiness) – this really is strength through (a special type of) weakness.

> It is unfortunate that the words stress and strain start with the same three letters and in common language mean the same thing, but we are stuck with the terms and you just have to get used to remembering which is which.

We often have to discuss work of adhesion and surface energy, each of which is measured as Joules per square metre, $J\,m^{-2}$. It happens that work is the same as energy, which is why the two measures have the same units and why "work of adhesion" is sometimes called "energy of adhesion".

The strength of an adhesive joint is often measured in force per unit length, i.e. $N\,m^{-1}$. If you sit down and do the sums (or if you go to my app page https://www.stevenabbott.co.uk/practical-adhesion/basics.php) you will find that a peel strength of 1 $N\,m^{-1}$ is the same as a work of adhesion of 1 $J\,m^{-2}$ (Figure 2.6).

It is tricky to know whether to add a space between a number and a unit. Some prefer $1N\,m^{-1}$, others prefer 1 $N\,m^{-1}$. For clarity and readability I have standardized on using the space.

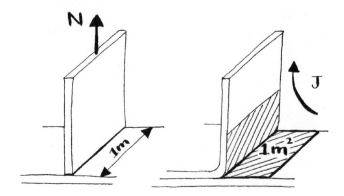

Figure 2.6   Classically, adhesion is measured as the force, N, per width, m in
$N\,m^{-1}$, or as work, J, to create 1 m^2 of separated adhesive, in $J\,m^{-2}$.
The two measures are exactly the same!

2.4   ADHESION TERMS

The word for the stuff doing the sticking is "glue" or "adhesive".
What about the word for the thing being stuck? "Adherend" is
the technical term, which can lead to phrases such as "... the
adhesive attaches to the adherend..." This isn't particularly
elegant, but there is no obviously better term to use.

We often have to discuss adhesion in terms of peel (pulling
vertically) and shear (pulling horizontally). There is a different
sort of vertical pull, the butt pull. Figure 2.7 gives you a visual
indication of what these terms mean.

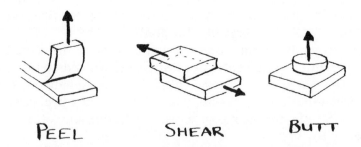

Figure 2.7   Whether a joint is tested in peel (vertical forces), shear (horizontal
forces), or butt (whole sample vertical pull) makes a great difference to the effective strength.

2.5 SURFACE TENSION AND ENERGY

If you go into your kitchen, take a random assortment of smooth, flat surfaces and place drops of either water, water with a dash of dishwashing liquid, or oil onto the surfaces you will see that the drops form different shapes, depending on both the liquid and the surface.

The liquids have different surface tensions, that is, different forces pulling the liquid together at the surface. Why are there forces pulling the liquid together at the surface? Molecules in the liquid are attracted to each other (if they weren't the liquid would instantly vaporize). At the surface of a drop this self-attraction becomes visible because the way to maximize their self-attraction (or minimize the number of missing attractions at the surface) is to form the smallest-possible surface which is a sphere. Drops containing dishwashing liquid have much weaker attractions at the surface because the surface is covered by a monolayer of the surfactant/detergent with long hydrocarbon tails that interact rather weakly; similarly, the oil has a low surface tension because it is made up mostly of long tails.

The surfaces have different surface energies. This arises for the same reasons – the molecules at the surface want to be with each other to a greater or lesser extent.

If the molecules at the solid surface have a large surface energy (for example a metal surface) compared to the drop then the drop will prefer to spread out on the surface, gaining maximum surface-liquid interaction, rather than curl up on itself in air. On a surface with a low surface energy, such as a Teflon frying pan, the same drop will spread out far less. Oil has a low surface tension and will spread out more on any of your kitchen surfaces than a drop of water, though the difference will depend on the surface itself. The water with the detergent also has a low surface tension so it too will spread out easily on many surfaces – one of the reasons we add detergents.

If you were to look carefully at each drop you would find that it makes a specific angle, called the contact angle, which is diagnostic of the relative surface tensions and surface energies (Figure 2.8).

If you have a Teflon frying pan, then this is "hydrophobic" (water hating) so a water drop does not spread. Even oil is un-impressed by Teflon's surface energy and oil drops hardly

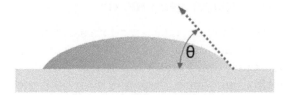

Figure 2.8   A drop of liquid forms a shape that depends on a combination of the surface tension and surface energy. The result is that the liquid meets the surface at a contact angle, θ.

spread. With some tricks it is possible to make a "super-hydrophobic" surface where a water drop is so unimpressed by the surface that it will roll along at the slightest tilt. These superhydrophobic surfaces have attracted a lot of excitement as self-cleaning surfaces but, although they are marvellous as lab demos, they are generally unsuited to the real world.

It is a common myth that surface energy is important for adhesion. As we will learn in Chapter 3, it is irrelevant for any sort of strong adhesive bond and is used only for those cases where easy breaking of the bond is a requirement – such as when a gecko needs to walk, or you need some temporary adhesion for wrapping a sandwich in a thin plastic film.

2.6   VISCOSITY

We are familiar with the fact that some adhesives are thin and runny while others are thick and resistant to flow. A thick adhesive can be said to be "viscous" and the measure of the ease of flow is "viscosity". If we define water as having a viscosity of 1 (the units are cP, which means centipoise) then some typical numbers for familiar materials are found in Table 2.1.

Table 2.1   Typical viscosity measurements of familiar materials.

Liquid	Viscosity, cP
Water	1
Olive oil	100
Glycerine	1000
Honey	5000
Ketchup	50 000
Lard	100 000
Peanut butter	250 000

Adhesives manufacturers have to worry incessantly about viscosity. Users can be equally as frustrated by an adhesive that is too viscous to flow nicely into a joint, as by one with such a low viscosity that it runs everywhere. Fortunately, there are some tricks they can play. For many viscous adhesives, if you apply a shearing or sliding force, their viscosity decreases rapidly (this is called "shear thinning"), allowing them to flow in the gap between the substrates. Others are "thixotropic" which means that they become thin when stirred or sheared (so they are also shear thinning), making them easy to apply, then, after some time, become thick again, stopping them from "slumping" out of a joint.

The study of how things flow is called rheology and interested readers can find a friendly (but technical) eBook and set of apps on my Practical Rheology web pages, https://www.stevenabbott.co.uk/practical-rheology/.

Even if the manufacturer has created the perfect viscosity, there is still the problem that squeezing a joint together is not as simple as it sounds.

2.7 SQUEEZING GLUE

It may seem odd to be discussing the simple act of squeezing some glue between two surfaces. What can possibly be of interest in such a basic process? It turns out that many of our frustrations and problems arise because squeezing is far more difficult to control than we might imagine. There are times when we need a thick layer of adhesive. Other times we need as thin a layer as possible. It would be nice to put one drop or blob in the middle and squeeze it evenly to the edge – a large drop for a thick layer and a small drop for a thin one. Unfortunately in the thick case it is far too easy to over-squeeze and get an excess of glue that is messy to wipe off because, well, it's rather sticky. In the thin case it can prove too difficult to get the adhesive to reach the edge at all.

The frustration arises thanks to Stefan's law of squeezing (Figure 2.9). There are various ways of thinking about it, and interested readers can find out more in this app, https://www.stevenabbott.co.uk/practical-adhesion/drop-squeeze.php.

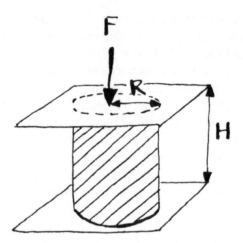

Figure 2.9   Stefan's squeeze law tells us that squeezing a cylinder of glue of radius *R* and height *H*, gets considerably harder as *R* increases and *H* decreases and it becomes near-impossible to obtain a very thin layer of glue.

Stefan's law tells us a number of things about how quickly the drop spreads and the thickness decreases:

- The higher the viscosity, the slower the process. This is intuitively obvious. A "gel" superglue with 20× the viscosity will spread 20× slower than a "pure" superglue without additives. This, with perhaps some shear thinning behaviour, is under the control of the adhesive manufacturer and with things like superglue we can choose which viscosity to use.
- As the drop expands it gets super-hard to expand further. If it takes 1 s to go from, say, 1 mm diameter to 2 mm, it will take 16 s to go from 2 mm to 4 mm. For those who like equations, the rate of increase of the radius, *R*, is proportional to $1/R^4$.
- As the thickness of the drop decreases it gets super-hard to decrease it further. If it takes 1 s to go from 0.4 mm to 0.2 mm, it will take 8 s to go from 0.2 mm to 0.1 mm. In equation terms, the rate of decrease of height, *H*, is proportional to H^3.

https://youtu.be/M0z1Xq52-HQ The video shows this nicely – six individual drops squeeze out to almost 3× the area of one blob containing six drops.

Because drops with a large thickness are easy to squeeze, if we start a little too thick then we can easily over-squeeze. Because

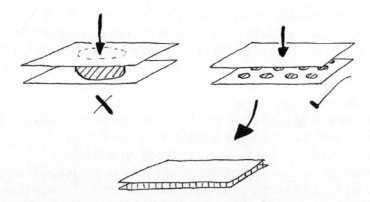

Figure 2.10 Getting around Stefan. Instead of one large drop, use a number of smaller ones.

thin drops are hard to squeeze, if we start with a thin drop in the middle then it becomes very difficult to get the drop to come right to the edge of the joint.

If you are in the unfortunate position of having a thick layer of adhesive at one side of a joint and a thin layer at the other then, well, give up. It is super-difficult to get such a layer to re-adjust itself.

If we combine the previous paragraphs with Stefan's law, especially the bit about the difficulty of expanding R, the least bad way to achieve a good overall thin coverage is to place lots of little drops across the surface (Figure 2.10). They can easily expand (because R is small) and become self-adjusting in terms of thickness (if H of one is small, a thicker H can more readily reduce its thickness). All the demos I've seen of superglue holding up some heavy truck from a crane start with a bunch of dots of glue (on a super-polished metal surface). Whether they know of Stefan or not, clearly the technique has been found to be reliable. The "dot and dab" technique for plasterboard/drywall, the blobs of cement for paving slabs, and the tile adhesive comb technique are variations of this theme.

2.8 GRIFFITH'S LAW

We all know that bits of junk can get in the way of good adhesion, so we like to ensure that surfaces are clean before sticking them together. One reason for this is described by Griffith's Law.

An otherwise strong joint can easily fail if the stresses at the interface encounter a weak zone, such as a bit of dirt or its opposite, a void from a bubble of air. Our instincts tell us, correctly, that while a very small particle or void is not a problem, a bigger one is more of problem. Griffith's law tells us that if the size of the void is *a*, the stress required for failure is proportional to $1/\sqrt{a}$. As the size of the defect gets *bigger*, the stress required for failure becomes *smaller*. When we make an adhesive joint, any defect from junk or air can be the source of a Griffith's crack failure, with larger defects leading to earlier failure.

As we shall see, when we try to break an adhesive joint apart, stresses are seldom distributed evenly across the joint. If a particle or void happens to be located in an area of low stress, it won't cause a problem. Conversely, if it is at a point of high stress, failure is assured. One of the frustrations of adhesion is that things seem to break for no obvious reason. Sometimes the reason might be a bit of otherwise harmless junk that's in the wrong place at the wrong time.

Griffith's law tells us that if we regularly attend to keeping adhesive free of large dirt particles and air pockets, our chances of unwelcome surprises reduce accordingly. That extra bit of cleaning before you assemble a joint really is worthwhile.

2.9   CHEMISTRY ESSENTIALS

We need just a few chemistry essentials to understand most of adhesion. Don't panic if you didn't enjoy learning chemistry.

The hydrocarbons which contain just Carbon, C, and Hydrogen, H, are rather neutral molecules that, in ordinary circumstances, contribute nothing to adhesion. The exception is when the hydrocarbons contain Carbon–Carbon double bonds, C=C, that can polymerize. In normal chemical drawings, the carbons are shown as simple lines. They actually have hydrogens attached to them but adding them to the images clutters things up unnecessarily. In Scheme 2.1 are the images of a few important groups, discussed below, to help visualize a few key ideas.

We are familiar with water as H_2O which we can show as H–OH, with one hydrogen atom attached to an oxygen + hydrogen group. I write it this way because much of adhesion depends on the availability of –OH (or "hydroxyl") groups where the oxygen plus hydrogen group is part of a larger molecule. If the –OH is attached to an ethyl group containing two carbons and five hydrogens, we

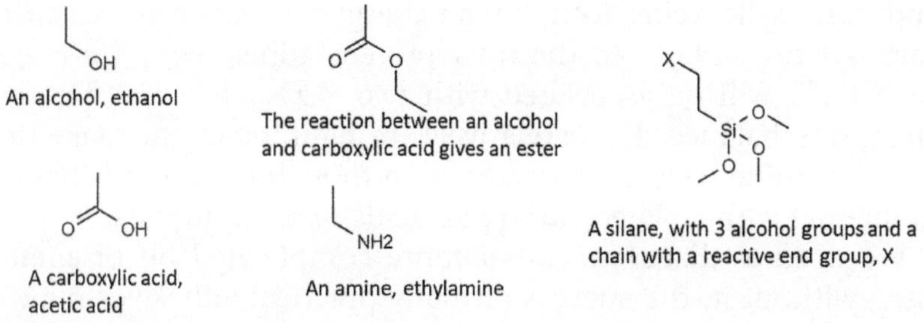

An alcohol, ethanol

The reaction between an alcohol
and carboxylic acid gives an ester

A carboxylic acid,
acetic acid

An amine, ethylamine

A silane, with 3 alcohol groups and a
chain with a reactive end group, X

Scheme 2.1

have ethanol, ordinary alcohol. There are two important things about –OH groups.

1. They like to associate with other –OH groups via a loose connection called a hydrogen bond. If a surface has some –OH groups and an adhesive also has some –OH groups then this loose association can encourage adhesion. However, water, with its H–OH can also associate just as well, so these loose associations are easily swamped by the presence of water.

2. They can react with a number of other groups to form stable chemical bonds that can increase adhesion. A carboxylic acid has a –CO_2H group which can react with an –OH to form a stable ester bond. The ester in the image is the result of ethanol reacting with acetic acid. The –OH groups can also react with silanes, as discussed shortly.

Nitrogen atoms can form –NH_2 groups ("amines") which, like –OH groups, can self-associate and also react with things like carboxylic acids. They especially like to react with the key components of epoxy and urethane adhesives as well as the acrylates in ultra-violet (UV) adhesives.

In addition to forming full chemical bonds, the carboxylic acids with their –CO_2H groups are also happy to form relatively strong associations with surfaces containing –OH and –NH_2 groups as well as with metal ions. As we saw in Chapter 1, adhesives based on milk proteins work well in the presence of calcium (Ca) ions which interact strongly with the numerous –CO_2H groups within the protein. Ions are the charged form of atoms and molecules. Calcium ions have two positive charges and are shown as Ca^{2+}

and carboxylic acids form mono-charged negative ions called carboxylates, $-CO_2^-$. In the milk protein adhesives, on average each Ca^{2+} will be associated with two $-CO_2^-$ in order for the charges to balance. These relatively strong interactions make the adhesive quite resistant to water, even though the water is happy to interact with isolated carboxylic acids or calcium ions.

We need to discuss a rather more complicated bit of chemistry; without it, the success of many practical adhesives makes no sense. The complication is not so much the chemistry itself, but the fact that we have to distinguish between different "Sil"s:

- Silicon, the element, on which we make silicon chips, and which can be reacted with oxygen and other molecules to create:
  - Silica (the mineral)
  - Silicates (more minerals)
  - Silicone release paper to which nothing sticks
  - Silicones used in cosmetics for their slippery feel
  - Silicones or siloxanes (bathroom sealants) which stick well, and
  - Silanes

Each of these reacted forms of silicon has the equivalent of four chemical bonds around them. For a silane, one bond is to a carbon atom (which itself is part of a medium-size molecule or a polymer chain), and the other three are to alcohol groups such as methanol or ethanol. The silanes are important because they are happy to swap one or more of the alcohol groups with $-OH$ groups from, say, the surface to which we want to stick something. This means that we can get strong bonds to any $-OH$ containing surface, using an adhesive that contains silanes. We can also start by reacting a silane with a surface and use the group at the other end (for example an amine) to react into the rest of the system. When silanes react with themselves this produces the polymerization ("curing") of the siloxane adhesives. Finally, the silanes are part of the curing system of the modern "silane hybrid" adhesives and sealants which are short, ordinary polymer chains stuck together via these silane reactions.

The reason that silanes are so important is that they can react with, and provide adhesion to, so many surfaces. Ordinary materials such as bricks or cements are themselves made of silicates which have –OH groups able to react with silanes. Similarly, although glass can be thought of as being relatively unreactive silica, there can be plenty of silicate groups at the surface. At the same time, most metal surfaces have a thin layer of "oxide" at the surface. Sticking to aluminium, Al, is not sticking to the metal but to an "aluminium oxide" surface. which loves to react with silanes to create "alumino-silicates". However, "oxide" surfaces can be complex and many of them do not contain many free –OH groups. This means that for many systems the surface has to be treated so that it contains enough –OH groups to react with silanes.

How can silicones at the same time be super slippery as in cosmetics and release papers, and super-good adhesives? If you have pure siloxanes, with no silane groups, then they are remarkably wriggly materials that can flex and twist very easily, giving them their special slippery feel and release properties. As we will discover in Chapter 4, it needs just a few reactive groups (the silanes) in the right place to convert a system to strong adhesion, making it possible to convert the slippery molecules into good adhesives. Silicone systems can also be made rigid by adding yet another variant of silicon, the delightfully named cube-shaped silsesquioxanes.

There is one more variant of the silicon-based adhesives. Because silicates are the basis of many rocks, why not make a rock-like adhesive from silicates? Sodium silicate solutions can indeed act as adhesives in special applications, though the upside of their strength comes with the downside of being brittle.

2.10 POLYMERS

Many adhesives are polymers and a lot of adhesion is to polymers. To avoid tedium, I will sometimes use abbreviations for many of the common polymers.

- EVA   (Poly)Ethylene vinyl acetate
- EVOH   (Poly)Ethylene vinyl alcohol
- PE   Polyethylene, commonly called polythene
- HDPE/LDPE   High Density PE and Low Density PE
- PP   Polypropylene
- PMMA   Polymethylmethacrylate, commonly called
              Perspex or Plexiglas

- PC   Polycarbonate
- PET   Polyethyleneterephthalate, commonly called
                    polyester
- PDMS   Polydimethylsiloxane, commonly called silicone
- PTFE   Polytetrafluoroethylene, commonly called Teflon
- PVOH   Polyvinylalcohol
- PVA   Polyvinylacetate; confusingly PVA can also mean
                    polyvinylalcohol
- PVB   Polyvinylbutyral
- PVP   Polyvinylpyrrolidone

We cannot understand adhesion and, especially, the trade-offs between strength, hardness and toughness without understanding a few polymer basics.

The simplest polymers are "vinyl" molecules, i.e. they are formed from those carbon–carbon double bonds mentioned earlier (the double bond is denoted in text by an $=$ sign, so: C=C), to create a string of carbon–carbon single bonds (denoted by a dash '–', so: C–C). As mentioned above, by convention the drawings do not include the hydrogen atoms that are also attached to the carbon atoms. In another convention, other groups attached to the carbon–carbon double bond can be denoted by the letter 'R'.

Let's first make some polyethylene. We start with ethylene, C=CR, where R happens to be hydrogen. We react it with a molecule called a "radical". We will show (Figure 2.11) it as X$^{\bullet}$

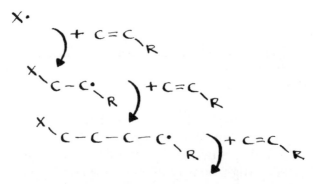

Figure 2.11 A radical polymerization of a vinyl molecule where R can be hydrogen (making polyethylene) or one of many other groups described below. It starts with one radical and carries on because each addition shifts the radical to the end of the growing molecule, ready to react once more.

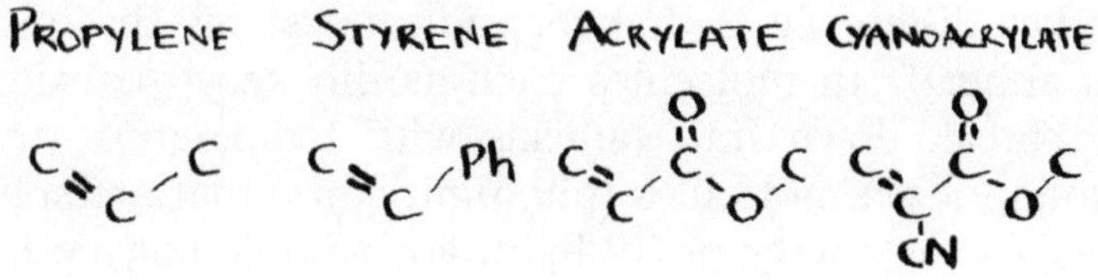

PROPYLENE   STYRENE   ACRYLATE   CYANOACRYLATE

Figure 2.12 A few vinyl groups, resulting in polypropylene, polystyrene, poly-acrylate or polycyanoacrylate.

where the dot indicates that the molecule is keen to react. Where the first X• (an "initiator") comes from will be discussed later. If we add X• to C=C we get X–C–C•. The dot on the X has reacted with one of the carbons, leaving a new free dot (radical) on the other carbon atom. This can react with another ethylene molecule to create X–C–C–C• and the process continues to create the long polymer chain.

There is a large choice of vinyl molecules, each with different Rs or with more than one R around the double bond (Figure 2.12).

Naming is not consistent, and so vinyl groups that constitute styrene give us polystyrene (rather than polyvinylbenzene) and groups that constitute acrylates are called polyacrylates or acrylics. A vinyl group containing a cyanide *and* an acrylate is a cyanoacrylate, giving polycyanoacrylates or superglues. These are especially good at polymerizing via a negative charge (an anion) (Figure 2.13) where you simply substitute C–C⁻ for C–C• and the process is otherwise similar.

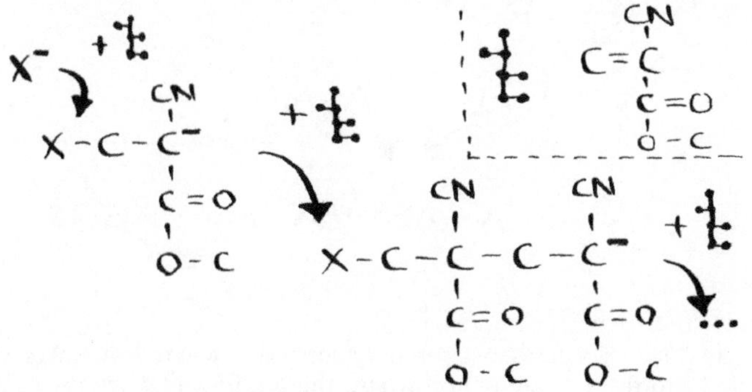

Figure 2.13 When we use superglue, the polymerization of the cyanoacrylate proceeds *via* a series of anions (negative charges) instead of radicals.

The key difference is that for radical systems the initiation comes either from molecules such as dibenzoyl peroxide that like to break down into radicals with temperature (thermal initiators), or via molecules (photoinitiators) that are split into radicals via the energy of UV light. For anionic polymerization, we need a "base" (the opposite to an acid) which can be a negatively charged HO^- group from water or an amine group. Common "accelerators" for cyanoacrylates are water, water made basic with sodium hydroxide or sodium bicarbonate, or a solution of an amine. The cyanoacrylates can also polymerize via radicals, so tubes of glue often contain a radical inhibitor to avoid the risk of them going solid before the tube has been opened for the first time by the user.

A different type of polymerization occurs when molecules are set up to react repeatedly at each end to create a polymer chain. One common type of reaction, mentioned earlier, mixes an acid with an alcohol to create an ester. If you mix a di-acid with a di-alcohol, the "condensation polymerization" can carry on indefinitely, to create a polyester (Figure 2.14).

Or if you start with a cyclic molecule called an epoxy, an alcohol group can react with one side of the triangle, creating a new alcohol group that can further react (Figure 2.15).

This is one form of epoxy adhesive. If a di-epoxy is mixed with a molecule containing a di-amine then these react rapidly to form a polymer. A dual tube epoxy has the di-epoxy in one tube and the di-amine (plus some tri-amine to create crosslinks) in the other.

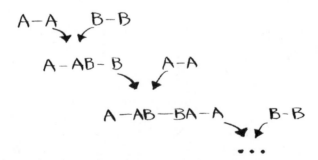

Figure 2.14   This is a condensation polymerization where A–A reacts with B–B to from A–AB–B, and where the A and B end groups of the new molecule are free to carry on reacting.

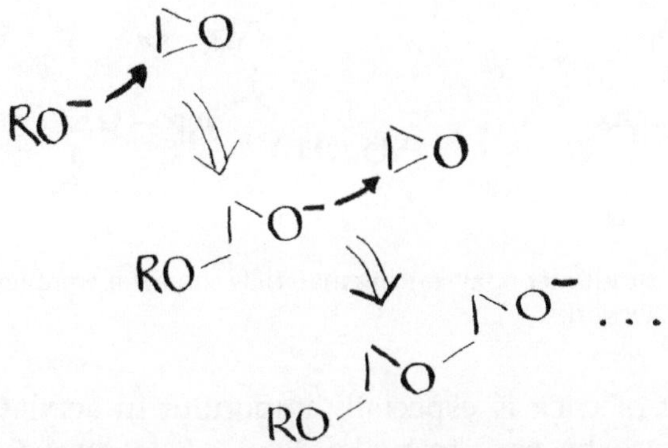

RO⁻

RO⁻

RO⁻ ...

RO

Figure 2.15 The anion RO⁻ reacts with the triangular epoxy, opening the ring to create another O⁻, which can react further. In epoxy adhesives the epoxy rings react with amines, a much faster reaction.

2.11 CROSSLINKED POLYMERS

So much for standard linear polymers. They have many desirable properties but often they are not rigid or hard enough for a specific adhesive function. To add rigidity, we make sure that the polymer chains form bonds, "crosslinks" between themselves, hence we call them crosslinked polymers (Figure 2.16).

To create a crosslinked polymer, whichever polymerization system we choose, the key is to make sure that the basic units to be polymerized contain extra functionality. If a standard unit is termed 1-functional, we can make 2-, 3-, 4-functional equivalents. A typical epoxy system might start with a 2-functional epoxy and a 3-functional amine (Figure 2.17). This will create a somewhat complex network. A 3-functional epoxy and a 4-functional amine would create an even more complex network.

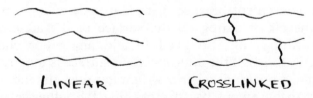

LINEAR   CROSSLINKED

Figure 2.16 In a linear polymer, each chain is, in principle, independent (in practice they tangle like spaghetti). In a crosslinked polymer, chains are chemically linked to each other, giving very different properties.

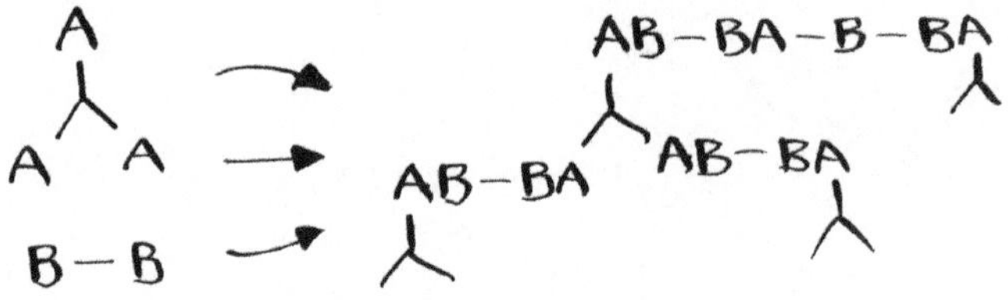

Figure 2.17   Multifunctional starting materials can form complex crosslinked networks.

This sort of trick is especially important in acrylate systems. It happens to be easy to make 2-, 3-, 4-, 5- and 6-functional acrylates. We will see that mixtures of these systems are the key to many adhesive systems, from nail polish to dental fillings.

For these crosslinking processes we often say that the system "cures" (or "sets") rather than "polymerizes". It makes no difference which word is used: the process of setting solid or becoming fully cured is a polymerization.

2.12   WATER AS THE INITIATOR

The beauty of polymerization is that you just need to get it going and then it carries on, in principle, until all the monomer is used up. In practice, polymerizations terminate when they meet impurities or another reacting chain, so we need a moderate number of initiators to get a good conversion. For superglues, urethane glues and silanes the first step of the polymerization reaction can be initiated by water molecules. This is wonderfully convenient and is an element of the popularity of these types of adhesive. It is also frustratingly inconvenient.

- It is convenient because there is usually enough moisture in the air or sitting on the surface of the adherends to start things off. No need to add any other chemicals ("activators"). If the atmosphere or surface is too dry, a huff of moist breath or a spritz with water does the job.
- It is inconvenient because opening the adhesive package to squeeze out the adhesive allows at least a few water molecules to get into the bulk of the adhesive, which will eventually go solid.

I always have problems with those tubes of silicone adhesives that are applied with a squeeze gun. The tubes are large enough for many jobs, yet I tend to need rather little for any single job, and the jobs are infrequent. For the next job I usually find that some portion of the silicone has gone hard; if I take off the cap to remove a solid blob from within the tube, I can do that job, but have added so much moisture that the tube is fully solid the next time I need it.

2.13 CATALYSTS

Many polymerization reactions happen rapidly once the reactants mix or an initiating radical is created by heat or light. Some, however, are rather slow. This can be good for those jobs that require a lot of fiddling to put things properly into place. When we want such reactions to go faster, we add a catalyst. Catalysts are molecules that get involved in the polymerization reaction, greatly speeding it up. But after each step, the catalyst emerges unchanged, ready to help speed up the next step. A small amount of catalyst, therefore, can have a large effect on the rate of polymerization.

Sometimes catalysts can be simple molecules like acids – which might be called "accelerators" rather than catalysts. Sometimes the systems require more sophisticated molecules. For example, the polymerization reactions for urethanes (general adhesives) and silicones (e.g. bathroom sealants) are catalyzed by tin-containing molecules. A problem with some catalysts is that they contain elements (such as tin) which carry small health or environmental risks so there is always pressure to reduce or eliminate them. Because catalysts are needed only at low levels, in general the risks from them are small.

For professional workers, fast cures are desirable. For amateurs, speed can be a problem. Different versions of the same adhesive might use different levels of catalyst for different speeds. Or an "anti-catalyst" (retardant) might be added to slow things further.

2.14 THAT'S IT

One more thing. Like most people, I had gone through much of my life tending to focus on the adhesive. During a very heated debate (scientists can get quite passionate) with some top scientists at a large chemical company I felt the frustration of positions hardening and drifting ever further apart, until one of their scientists, Dr Michele Seitz, said, "But adhesion is a property of the system". That one sentence entirely changed the debate and it turned out that both sides had been focussing too much on specifics and not on the system. It also entirely changed my approach to adhesion, and I remain grateful to Dr Seitz for her momentous intervention.

Now we have all the basics in place, it is time to start exploring adhesion. And a good place to start is without any adhesives at all. In other words, it is time to learn about geckos.

CHAPTER 3

Sticking Like a Gecko

Congratulations! You have just been promoted to chief science officer of the geckos. And you have been presented with geckodom's biggest challenge yet. How are you going to climb up all those smooth glass structures that humans have created? You know that the smooth surface gives you nothing with which to grip. What are you going to do?

Some of your fellow geckos suggest that you create a glue gun for your lizard feet, but you immediately reject this idea. "The problem with glue isn't how we are going to stick, but how we can unstick ourselves. A glue will be too good – we'll hang there, unable to move". Translating this into scientific language you say: "What we want is the world's worst adhesive – just strong enough to hold a gecko, yet weak enough to be easily broken when we need to take another step".

Alone in your well-equipped lab, you look with annoyance at some dust on your equipment. It is *everywhere*, even on the sides. This gives you an idea. You get out your atomic force microscope (AFM) and attach a bit of dust to the end of the probe ready to measure the force between the dust particle and the surface. What you want to do is push the particle onto the surface and measure the force needed to pull it off, yet to your surprise, as the particle gets close to the surface it jumps into contact (Figure 3.1). It really wants to be on that surface!

Sticking Together: The Science of Adhesion
By Steven Abbott
© Steven Abbott 2020
Published by the Royal Society of Chemistry, www.rsc.org

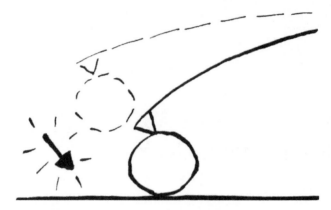

Figure 3.1   A bit of dust on the end of an AFM tip is pulled into contact with
the surface. This is a surface energy effect.

At first you think this is because the particle is statically charged,
but even when you carefully neutralize the static with a deionizer,
the particle still jumps into contact. You try many different types of
dust and many different types of surfaces and you find that the
attractive force does not change much. It seems to be a general
property of all materials that they attract each other with a small
force. [This general attraction is called the van der Waals force after
the scientist who first characterized it properly.]

As a good scientist you want to give this force a value. The force on
its own doesn't tell you much because it depends on the area of
contact. To make the value universal you decide to look at it in terms
of the amount of energy in Joules, J, needed to separate one square
metre of surface, and call it surface energy in units of $J\,m^{-2}$. For all
surfaces you can find this varies from a low of 0.02 to a high of 0.06,
with most being around $0.04\,J\,m^{-2}$ or, for convenience, $40\,mJ\,m^{-2}$.

When you then do the surprisingly complicated calculations
about how much of a gecko's weight could be supported if you
only had those surface energies, you find a clear answer. If the
whole area of your four feet was in contact with the smooth glass,
you would never move again – it could support 100 kg. You
therefore need only a modest fraction of your feet to be in perfect
contact. Examining your feet you realize that they have a complex, multi-level design that allows lots of contact (Figure 3.2).
For the simpler feet found on most other animals, the total
contact area with smooth glass would be so small that there is no

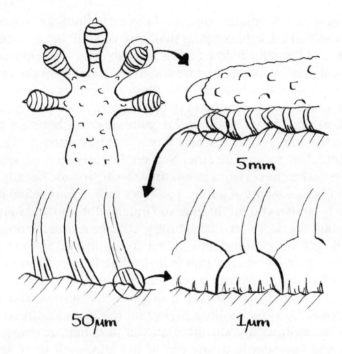

Figure 3.2 The hierarchal structure of a gecko foot giving compliance to the surface at every scale from mm down to the sub-nm structures visible in the 1 μm view.

hope of holding on. The problem is that for surface adhesion you need "contact" and this means being within 1 nm of the surface. Any "normal" foot is rough to at least the 1 μm level and usually the 1 mm level, so the total area in contact with the glass would be far too small to provide grip.

But your feet seem amazingly well designed. You have toes that can make sure that each pad of the foot gets close to the glass, then you have lamellae (5 mm image) on your toe pads that can adjust into broad, good contact, then you have setae (50 μm image) on the lamellae that can adjust into fine scale contact and then the setae have spatulas (1 μm image) that come into intimate nano contact. Indeed, the spatulas are remarkably like the cantilevers used in AFMs to allow nice, controlled contact with any surface. You realize that your whole system is compliant to the surface – able to accommodate to its ups and downs at every relevant scale.

You are a scientist, you have done your calculations; it is time to make your announcement. "My fellow geckos, tomorrow I will show how geckodom can conquer the glass buildings of mankind. Meet me at the local shopping mall and you will be amazed".

The next day your fellow geckos assemble and you clamber up some easy wall to reach the glass. With confidence in the laws of physics you put one foot onto the glass and expect to feel a solid grip. You test it and a moment of panic arrives – there was almost no adhesion. You try again and to your surprise there is a good, solid adhesion. You try the next foot – your first attempt is a near disaster, then things are fine. You confidently step on with all four feet. The cheers from below are starting to fade because you hit another moment of panic. You can't move. Your calculations warned you that you might have too much adhesion and here you are, totally stuck. To buy some time you make a speech about one giant step for geckokind and then a cramp in one foot creates a muscle spasm – and your foot is immediately freed. You vaguely recall a lecture on fracture mechanics and make a wild guess that somehow sudden motion could overcome adhesion. You place your newly-freed foot a little higher on the glass and then try a "flick" motion with your ankle on the next foot. It comes free easily and you scuttle to the top of the glass wall then, with a gasp from the geckos below, you even walk upside down on the glass canopy.

3.1 LESSONS FROM THE GECKO

That little tale conceals a rather advanced master class in adhesion science. Let's review it from a human perspective:

- You now know that everything is attracted to everything else with a rather modest force named after van der Waals.
- This van der Waals (vdW) force varies rather little between materials and gives us, for materials we are likely to encounter, surface energy values in the range of ~ 20 mJ m^{-2} for "low surface energy" materials like silicones or Teflon through ~ 30 for plastics like polythene, though ~ 40 for many other plastics and for glass, then up to ~ 50 for real-world "metal" surfaces which are generally covered with some sort of oxide. This immediately tells us that for the sort of strong adhesion most of us need, surface energy is irrelevant. If the whole range of practical materials changes in surface energy by a

factor of 2.5, it cannot possibly explain how adhesion can vary by factors of 1000 between different surfaces and adhesives.

- *If* (and this is a big if) you can get near-perfect contact between two surfaces, then the effective adhesion against a simple pulling force can be rather large. The classic example is when two beautifully machined metal surfaces are put into contact. They are impossible to *pull* apart, but rather easy to slide apart. Their adhesion, measured in one manner, is large, yet their general purpose practical adhesion is small. You wouldn't stick an aircraft together via pure surface contact!

- *If* (and this is a big if) you create two beautifully machined metal surfaces out in space where there is no oxygen to corrode the surfaces and if you can push them together then you don't just get strong adhesion, you get a solid piece of metal. Because metals are ductile, slight mismatches between the surfaces will flow out. The atoms that met across the surface have no way to know that they were from different surfaces, so the result is a perfect piece of metal. If, back on Earth, you can smash two pieces of metal together with sufficient force to push the contaminated surfaces out of the way, leaving pure metal-to-metal contact, then you have a "cold weld" which, in theory, is indistinguishable from a pure piece of metal. Although we have merely created pure surface contact, I think it is fair to say that we now have metal-to-metal *bonds* because there is a continuous electronic structure. This is not the same as having pure surface contact between molecules (such as polymers), where there is no continuous electronic structure.

- *If* (and this is a big if) you can place a drop of sodium hydroxide solution onto one super-flat piece of glass then place another piece in near-perfect contact, the solution etches the surfaces of both pieces of glass, gradually forming a gel which then solidifies. This is Hydroxide Catalyzed Bonding and is wonderful if you have the time and patience to practice the technique in a super-clean environment and need, for example, to send some Gravity Probe optics into space. Is this gecko-style? Yes, because it's a trick for producing perfect contact between two smooth surfaces, no because it is using something a bit like an adhesive to do it.

- The gecko gains its high levels of surface contact via its multi-level compliance: legs, toes, lamellae, setae, spatulas.

Human attempts to achieve gecko-style adhesion are usually far less compliant (they might have 2 or 3 levels, not 5); they are therefore far less capable of dealing with the real world.

- This pure surface adhesion is very easily disrupted by the most humble piece of dirt. The gecko's first step onto the glass was not successful – the foot had brought up bits of dirt from the wall it had just climbed. The gecko has no magic for getting good adhesion between its foot and a smooth surface when there is a small bit of dust in the way. The magic is in what happened when the gecko tried again. The adhesion of the dirt to the glass turns out to be greater than the adhesion to the foot; lifting the foot and trying again provided a self-cleaning action, allowing good adhesion on the next attempt.

- Although it is relatively easy to measure vdW forces and surface energies it happens to be very difficult to take these numbers and calculate the load that a gecko spatula or toe could hold. There are many logical steps involved and they will not be covered in this book.

- Finally, gecko adhesion is easily disrupted by a little flick of its heel. That flick induces a crack at the interface. Once that crack starts, there is nothing much to stop it, so the foot comes away easily. The key message in this book is that Adhesion is a Property of the System. If the system is "gecko upside down on glass" then it will stick there till the end of time (yes, dead geckos stick as well as live ones, it has been tested). If the system is "gecko creating a crack along the interface" then the adhesion is very small.

It turns out that geckos cannot climb on Teflon, though they have no problem doing so on wet Teflon. There are competing explanations for this, with my favourite being that the tips of the spatulas become more compliant when wet, allowing the little extra adhesion necessary for climbing. The chemical acronym for Teflon is PTFE, with the 'F' indicating to us that it is a "fluoro" type of polymer. We shall see later that the easily removed "silicone release paper" on adhesive tapes cannot be replaced by "fluoro release paper", even though the two polymer types give us the same low surface energy; so low adhesion comes about via other mechanisms that we shall discuss later. I have not found any data of tests of geckos on silicone release surfaces but I suspect that they will have trouble climbing them, even when wet.

While we are at it, it is worth noting that the same applies for adhesion as well as non-adhesion: silicone adhesives are often wonderful, but there aren't many fluoro adhesives. Our "non-stick" frying pans, on the other hand, *are* covered with fluoropolymers. Again, whether things stick or not clearly depend on many factors other than surface energy.

One last thing about walking on glass like a gecko. I once had to give a lecture on adhesion and had worked out that with a couple of pieces of rubber attached to my hands I could cling, gecko-like to the window of the lecture theatre. My request to do this was turned down: "It's not that we're against science demos; it's just that we don't fancy the headline 'Mad professor falls from 14th floor window trying to be a gecko'".

3.2 WHY DON'T WE HAVE LOTS OF GECKO ADHESIVE TAPES? ACTUALLY, WE DO

When the first scientific studies of gecko adhesion revealed the simple elegance of the mechanism there was a rush of popular articles (they still appear from time to time) saying that soon we would have gecko tapes giving us "chemical free" adhesion. A moment's thought would have saved a lot of wasted research effort.

Most of us, most of the time, do *not* want an adhesive that is trivially broken with a crack or a pull in the right direction. Most of us, most of the time, do *not* have super-smooth, super-clean surfaces to stick to, so pure surface energy will not work very well.

https://youtu.be/kiFDCfR817Y An example of the dangers of this type of adhesion is the Two Strong Men video which I made many years ago, but is still striking. It shows that two strong men (we weren't acting, we were genuinely trying our hardest) cannot *pull* apart two smooth sheets of rubber. Yet a young girl can *peel* them apart with no effort.

The really clever part about the gecko's foot is the hierarchy of mechanisms to ensure good contact: toe, lamellae, setae, spatulas.

Any piece of gecko tape which lacked this hierarchy might be OK on glass but relatively useless on most non-smooth surfaces.

I once argued with a famous gecko adhesive scientist, saying that even if you could make a gecko tape (which happens to be a difficult challenge) it would be of little use. To my surprise he entirely agreed with me: "But we have found that there are some medical applications where they are willing to pay the high price because they require both the strengths and the limitations of our tape". I was delighted with his reply.

In general it is impossible to get good surface energy contact between surfaces, so gecko-style adhesion is relatively rare. One example is called, in my home, cling film; Wikipedia tells us that it is called variously "plastic wrap, cling film, shrink wrap, Saran wrap, cling wrap, food wrap, or pliofilm". These films of polyvinylchloride, polyvinylidene chloride or, more usually these days, PE, are super smooth and cling together via pure surface energy. You can easily get a model gecko (or, in my case, Spider-Man) to stick to a pane of glass via a little rectangle of cling film which in turn (this is cheating, of course) is stuck to the gecko's foot (Spider-Man's hand) via a cocktail stick (Figure 3.3).

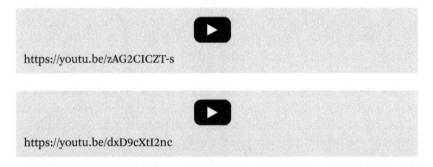

https://youtu.be/zAG2CICZT-s

https://youtu.be/dxD9cXtI2nc

Cling film provides a lot of science in a simple format. If you take a sheet and squeeze it carefully against another sheet, you find that the adhesion is quite impressive. But if you make the slightest mistake (as I usually do) and get a wrinkle, the effective adhesion reduces considerably. A bit of dirt or a crumb of food reduces the adhesion still further. In the clingfilm video we measure the peel strength from pure surface energy using cling film, though I would have got a more accurate number if I had eliminated the static that added some extra attraction.

Figure 3.3 Spider-Man clinging on with 10 cm² of clingfilm.

A rather more serious variant of cling film has a formal name "self-amalgamating tape" and a number of informal ones, including "F4 tape". Why F4? One of the more successful fighter jets in history was the F4 Phantom. Missions are very tough on aircraft and they would often need a temporary fix to get the plane back in the air quickly. Some cables might need to be bundled back together and insulated, or a leaking pipe might need to be fixed. For many of these jobs, duct tape was not the right thing (even if much of the rest of the plane was held together with it). The tape that did the job for the F4 (and many others, but the name remained) was a smooth, relatively soft rubber or silicone tape with a rough release liner. You wrapped the tape around the area of interest and, because it was smooth, it stuck together instantly. If you made a mistake you could easily peel it back and try again. It was easy to smooth out air bubbles and wrinkles. So far it is just a soft cling film. The trick was that the rubber or silicone had some low molecular weight polymer that could slowly migrate through the system, causing the boundaries to blur and to make adhesion perfect because there was no longer a crack to fail. Being soft it could withstand vibrations and the stresses and strains of active duty.

When it was time to do a full repair, the F4 tape, which had withstood tough combat missions, could be carefully cut along its length and easily peeled off.

Another example of surface energy adhesion that many of us use are some types of hooks that stick to relatively smooth surfaces such as glass, tiles or smooth paint and can then cope with loads of 1–2 kg yet are readily removed and re-stuck. Although they really can hold 1–2 kg, you need to pre-clean the surface carefully and have to attach heavy loads delicately, otherwise the sudden jolt creates a peel force. As we shall see, the resistance to the peel force (peeling away from the wall) is much lower than the shear (if the hook is on a wall) or butt pull (hook on the ceiling), so the peel allows you to easily remove and, in principle, reposition the hook. The problem with these gecko-style hooks is that the slightest bit of dirt between the two surfaces creates a crack path and most of the adhesion is lost. If they fail for any reason, such as a sudden shock or vibration, then they fail completely. If you read the feedback pages of a large online store you find equal numbers of "these are wonderful" and "these are terrible" stories, with the latter revealing examples of, say, holiday wreathes completely smashed when the hook's adhesion suddenly failed.

An alternative removable hook system is a variation on pressure sensitive adhesive tapes so is discussed in Chapter 6. Although the mechanism is very different, the online world is equally divided. When they work they work, when they fail they fail suddenly, which in a way is not surprising because they are *designed* to be easy to remove.

There is a further class of surface-energy-only adhesives that we throw away without a second thought, even though creating them is hugely difficult. These are the protective laminates we see on a new smartphone screen or many smooth shiny parts of other electronic devices such as TVs. Often, they are just a thin film of smooth, low density (meaning "not too crystalline") polyethylene. By being smooth they make good surface contact and the adhesion can be adequate. A standard low density PE is too crystalline to be able to adapt to small degrees of surface roughness, so its adhesion is too unreliable. By deliberately adding defects (carbon side chains) to the main polymer chain the PE can go from low density to ultra-low density which has very low crystallinity and is more forgiving to the surface, allowing higher adhesion. What if you need even higher adhesion? No problem: throw in a non-polymer that makes the polymer structure even weaker, providing even better contact with the surface, generating strong adhesion. The additive plasticizes the polymer, making it more flowable. Hopefully you can see the potential downside to these low-grade polymers with plasticizing additives: there is the risk of leaving a residue on the surface.

For these protective laminates, finding the right balance between good adhesion and leaving a residue is not so hard in a lab. The real world is much harsher. A protective laminate might sit on a surface for a year and during that time some of the junk in the polymer might accumulate at the surface, leaving a residue and an angry customer. It is a frustration for everyone who creates or requires protective laminates that the right balance of clarity, cost, adhesion and long-term resistance to leaving a residue is so hard to find. Customers never notice the 99% of the laminates that do their job perfectly and get irate with the 1% of laminates that have come loose, are too stuck to be removed easily, or have left a residue.

My own experiences wrestling with this issue have often concluded, reluctantly, that surface energy on its own is not sufficiently trustworthy. Despite the simplicity and low cost of these systems, the practical choice is a more expensive film (often clear PET which is always too crystalline to give reliable surface energy adhesion) coated with a low level, but reliable, pressure sensitive adhesive (PSA). Smartphone screen protectors which need to be secure (yet removable if desired) are stuck on with a very expensive, high quality PSA.

Writing a book is an adventure in challenging one's own assumptions and statements. In the same week that I wrote the original draft of this chapter I had two demonstrations that my earlier assertion that "In general it is impossible to get good surface energy contact..." was wrong.

First, a friend observed that items on his painted windowsill became stuck over time. He showed me an example where a picture frame was indeed slightly stuck to the paint. With a bit of peel motion it was easily removed, and there was no visible mark on the paint. Clearly the paint surface had flowed on the micron scale over weeks or months to create significant surface contact. He had never had this problem with old-fashioned solvent-based paints. As we shall see in Chapter 8, water-based paints need additives to help the paint particles to coalesce into a film. My guess is that these additives are present at a level sufficient to keep the particles slightly mobile, hence their slow flow into contact with the picture frame.

Second, I found myself having to do some work on a "cold seal" adhesive system. I had never thought about cold seal even though there is a well-known example, envelope adhesives which do not stick (much) to anything else but readily stick to themselves when pressure is applied. It turns out that these adhesives are similar to normal adhesive tapes but they don't create quick tack with another surface. The trick is to get them to produce adequate adhesion when pressed hard together, allowing the polymers to flow sufficiently to create the same type of adhesion as is generated spontaneously with a standard adhesive tape. The adhesion itself is not due to surface energy – these seals work via the same principles as pressure sensitive adhesives. By being on the borderline between pure surface energy and pressure-induced flow, they are an interesting example of surface energy at the margins.

3.3   LIQUID GECKO ADHESIVE

To get good gecko-style adhesion, all we need is perfect surface-to-surface contact. This is impossible to achieve (even for geckos) outside highly controlled laboratory conditions. When we start exploring common household adhesive tapes in Chapter 6, we will find that they must meet the "Dahlquist criterion" which states that the adhesive must be relatively soft and squidgy so that it can spontaneously flow into perfect contact. If it is not soft enough it will not give perfect contact with the surface. A PSA needs the good contact to provide full surface energy adhesion which in turn gets amplified by the adhesive itself. The cold seal surfaces of the previous section do not have the required softness to create good, spontaneous contact with other surfaces, yet together can provide adequate contact and adhesion when enough force is applied.

It is even more impossible to get two real-world solid surfaces into perfect contact without extreme pressures that force them to flow (the cold welding mentioned above). This is why liquid glues were invented.

If you can squeeze a little bit of liquid between two surfaces, it will make perfect contact with both sides. Some readers might know enough about "wetting" to immediately disagree with this statement. After all, a drop of water on a piece of plastic does not spontaneously flow to cover the surface. Although this is true, a drop of water placed between two pieces of the same plastic will spontaneously fill the gap. We will return to this apparent puzzle later.

Let us be specific and say that we have two pieces of a broken cup where we are well able to match the two jagged edges at the large scale but where they are simply too jagged to match at the nm scale. The liquid spontaneously fills the gap and we squeeze reasonably hard so that excess liquid is pushed out. Now, via some magic, let the liquid solidify (Figure 3.4). We have four surfaces in perfect contact: upper cup surface to upper glue surface and lower glue surface to lower cup surface. As long as we don't get a crack going along the interface, we now have surface energy holding the cup together.

The specific modern example of this is superglue. The liquid is a cyanoacrylate which is stable in its plastic tube. As it is squeezed onto the piece of broken cup it picks up moisture from the air. As shown in Chapter 2, a water molecule turns a

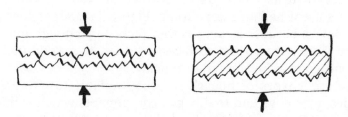

Figure 3.4 No practical pressure can force together the two solid surfaces, but a liquid can easily fill the gap. If the liquid turns to a solid then we have, at the very least, perfect surface energy contact between all four surfaces (two interfaces).

cyanoacrylate molecule into a cyanoacrylate ion which can readily react with another cyanoacrylate, creating an ionic dimer which reacts with another cyanoacrylate ... until (we hope) all the cyanoacrylates have been joined together.

We all agree that these superglues are amazing. Now ask yourself the question: "Do I trust this superglue fix on my favourite coffee (or tea) cup?" Experience tells us that superglues are wonderful for jobs where there aren't any sudden or oddly-angled stresses. A small jug for watering orchids that has been fixed with glue will stay together no problem. We can pick it up, fill it with water, put it down with a bit of a bump and have no fear of it falling apart. But would you trust a glued handle on your hot cup? I wouldn't. Just pouring in the hot drink can create thermal shocks that might break the joint. And it just needs one careless motion to impose an unexpected stress on the handle with a heavy load of hot drink for the joint to break and for your drink to spill all over you.

I am so old that I remember the superglue revolution. Before that we always had a problem. The only way to make a liquid adhesive was via a solvent (water, toluene, acetone ...). If you put the joint together too early for something non-porous such as a cup handle, the solvent had not evaporated so there was no viable adhesion. If you waited too long for the solvent to evaporate then the adhesive was too solid to give the intimate contact needed for surface energy adhesion. By getting it just right you had adequate adhesion which improved over, say, 24 hours as the remaining solvent molecules escaped along the edges of the joint.

Those solvent-based glues worked wonderfully on porous surfaces such as cloth, paper, leather or wood, because any excess solvent had an escape route (Figure 3.5). Superglues were a revolution because as long as they picked up enough moisture before you squeezed the surfaces together, they solidified and you had good adhesion rather quickly.

The other types of household glue available to us were the epoxies, where we had to mix two components which solidified over time. Because epoxies have the potential for an extra and highly significant degree of adhesion, we will discuss them in detail later in the book. For the moment we will take a simple

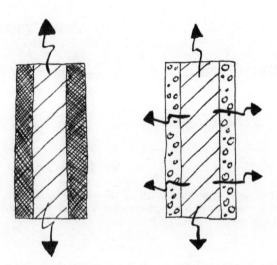

Figure 3.5 The problem with solvent-based glues is the solvent has to go somewhere. For impermeable surfaces the solvent escapes (too) slowly through the edges. For permeable surfaces such as leather the solvent can easily escape. Superglue goes solid without solvent evaporation so works for both systems.

case that highlights the problem with this sort of surface energy-only adhesion.

During the writing of this book I happened to need to use a nail brush to get rid of some glue on my fingers from a failed experiment. My hands and the brush were wet and suddenly the brush's handle fell off. I'd forgotten that I had broken it some years ago and had glued it back on with some superglue. It had been fine for years. Now the combination of a bit of pressure in the wrong direction and, I assume, the water meant that the handle came off. This then reminded me of a beautiful experiment described by one of the Big Names of adhesion science, Prof. Kevin Kendall in his famous Molecular Adhesion book.

He coated a thin layer of epoxy onto glass and, in his words, "... *the adhesion was good. The film required considerable force to wedge it off the glass*". He then placed a repeat sample of epoxy-on-glass into some water. The epoxy film floated off! The explanation is partly to do with surface energies and partly to do with water and glass. First, the surface energies.

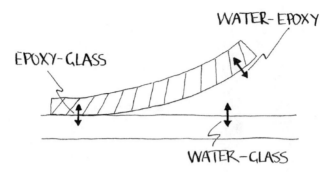

Figure 3.6   How water can destroy strong adhesion – by creating water-epoxy and water-glass interactions that in total are stronger than the original epoxy-glass interaction.

The epoxy surface has a choice: contact with air; contact with glass; contact with water. The glass has a choice: contact with air; contact with epoxy; contact with water. When you calculate the competing choices, although glass and epoxy prefer being with each other compared to being with air (so dry adhesion is good), when water is around, the balance of choices shifts to them both preferring to be with the water: the epoxy floats off (Figure 3.6).

Although the choice is clear, why do the glass and epoxy get the chance to go with the water? How can the water get between them? Water is a very small molecule and it is the humbling experience of anyone who wants to keep water away from an interface that it can usually find a way to get there. It is a very destructive molecule for those who work with adhesives because not only does it change the surface energy balance it can also attack many of the chemical bonds and weaker "hydrogen bonding" interactions that help hold surfaces together. However, such processes take time. I assume that the handle came off my brush at that specific moment because water had slowly been getting to the interface over the years of use since I repaired it and a critical amount had, at last, arrived.

In the Kendall example, the epoxy floats off rather easily because water has an especial attraction to the silica at the surface of glass, converting it to a silicate, and the process can zip along the interface rather quickly. This effect of water at the surface of glass means, incidentally, that you cannot reliably leave two sheets of glass in contact when there is a moist atmosphere.

Over time the glass at the contact point is "dissolved" by the humidity, leaving an annoying semi-opaque layer at the surface which cannot easily be removed.

3.4 USE SPARINGLY

It always used to surprise me that tubes of household glue instructed us to use a thin layer. If the glue is doing the sticking, surely more is better. Logic then stepped in. Any manufacturer wants the user to use as much as possible. If they are saying "use sparingly" they must have a very good reason.

In turns out that there are two reasons. One reason must await another chapter. The second is simple: most of these adhesives are rather useless polymers. This is not an attempt to insult the adhesive makers. Adhesion is a property of the system and the overriding priority is to find an adhesive formulation that is sufficiently liquid in the tube, with a sufficiently long shelf life before going solid in the tube, and with a sufficiently fast drying or solidifying time – along with desirable properties in terms of health, safety, price, odour etc. when in use. The end result is that (excluding crosslinked systems such as epoxies, urethanes, urea/formaldehydes etc.) the polymers are nowhere near as strong as apparently simple polymers such as PE or PET, and even less strong than the ceramic of a cup. Yet they are able to do their job if applied properly, because adhesion isn't just about strength.

This weakness of the adhesive means that if you have a thick layer in the joint, failure might be within the adhesive itself – this is called "cohesive failure" (compared to "adhesive failure" when the adhesion fails (Figure 3.7)) and is deeply unpopular with users; they don't like it when it is obvious that the adhesive itself wasn't up to the job. When the layer is as thin as possible, then the strength of the polymer itself is less relevant and/or less obvious when a failure is observed.

In Chapter 5 I will describe a special adhesive polymer which is one of the most useless polymers on the planet with very few desirable properties. If you include 100 nm of this polymer in a special type of joint, the adhesion is very poor – the joint fails by cohesive failure within this useless polymer. If, instead, you include a 20–50 nm layer then, as I once spent a few challenging months proving, the joint is almost indestructible.

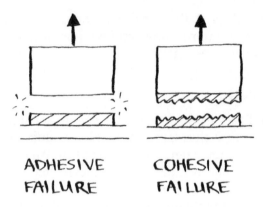

ADHESIVE   COHESIVE
FAILURE   FAILURE

Figure 3.7 The difference between adhesive failure (at the interface) and cohesive failure (within the adhesive).

3.5 THE PAIN OF CONTACT IN CONTACT LENSES

A downside of "thin" and a painful example of pure surface energy adhesion between nicely smooth surfaces is known to those who wear a pair of contact lenses rather longer than they should.

Under normal use, the lenses are held in place by surface tension forces from the liquid on the surface of the eye. You can readily get an idea of the scale of surface tension forces with a well-known hoop-like breakfast cereal. Float one hoop on the surface of your milk and with a pair of tweezers gently pull it upwards. You see the milk clinging to the hoop and you find it pulling a significant weight of liquid up from the surface till gravity exceeds surface tension and the hoop is freed. Those surface tension forces are not large, but for contact lenses they are more than enough to hold them in place. However, such forces alone can easily be overcome with a sudden shock, like blinking, so it requires an extra effect to keep lenses from popping out.

Remember Stefan's squeeze law? If you have a thin layer of liquid, it gets increasingly hard to squeeze it out as it gets thinner. Stefan's law works in reverse, too: if, instead of pushing, you try to pull two surfaces apart, it is hard for the liquid to flow inwards, at least for a while; you have to provide a significant force to allow that flow to happen at a significant speed. Fast, pull forces to the lenses are resisted by the difficulty of liquid flowing through a narrow gap. Together, this means that a normal moist eye has a contact lens floating on a thin and stable cushion of tears. Removing them is all about letting the tears flow easily under the lens as it is pulled away. Squeezing off a soft lens is easy. Blinking off a hard lens probably requires the blink to push liquid under the lens, to fight against the Stefan's law reluctance to flow.

The painful part is when the eye dries out after a long day of wearing lenses. Now you have surface energy contact with the eye. As we know from the gecko, these forces strongly resist a vertical pull and give a modest resistance to shear. If you could peel the lens then it would come off easily, but that isn't viable for most contact lens users: it is exceedingly tricky to catch the edge of a lens. So removing them is a difficult process of applying as much peel and shear (including the trick of holding the lens in place with a finger and rolling the eye away underneath it) and as little butt force as possible, all the while trying to induce tears to re-float the lens.

3.6 (DON'T) ROUGHEN BEFORE USE

One of the most enduring of adhesion myths is that you should roughen your surface before use to promote adhesion. We have already learned that this is wrong because surface energy adhesion is best when two smooth surfaces come together as intimately as possible. The more you roughen the interface, the harder it gets to ensure intimate contact. Yet the defence of roughening relies on surface energy; it says that if you roughen, you get a larger surface and therefore more surface energy adhesion. You also, they say, get mechanical interlocking between the surfaces, which sounds awesomely helpful.

Both the extra surface area and mechanical interlocking ideas are wrong in an interesting manner. Let's look at a typical rough surface measured by sliding a diamond stylus across the surface and measuring how its tip goes up and down (Figure 3.8). Although the trace in the figure is a simulation from one of my apps, it is realistic and instantly recognized by anyone who has measured a rough surface using this stylus technique.

That looks like (and in terms of typical surfaces, actually is) an amazingly rough surface and you can imagine how the adhesive

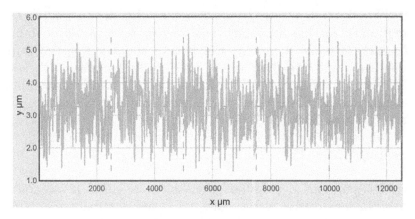

Figure 3.8   A typical output from a surface roughness measurement device. It looks an amazingly rough surface!

would love to cover all that vast extra surface area. It also suggests how you might create mechanical interlocking between surfaces – imagine the peaks of the top surface stuck down into the valleys of the bottom surface.

To work out how mountainous the surface is, imagine an ant that has to crawl along the smooth equivalent of this surface, along a straight line from left to right. It will crawl exactly 12.5 mm (12 500 µm), the standard scanning distance used for these probes. Now let it walk up and down those mountain peaks and valleys, using ant GPS to measure how far it drags its tired feet. Because I wrote the app that created that image I can look inside and get the exact value. It is 12 506.5 µm, meaning that the ant has only travelled 0.05% further.

We can make sense of this if we look more carefully at that roughness curve. It looks amazingly rough when everything is squashed up – 12 500 µm along the *x*-axis against 6 µm on the *y*-axis. As we gradually magnify things along the *x*-axis, to 1250 µm then to 125 µm and then to slightly more than 12.5 µm we see that the surface is really rather gentle (Figure 3.9).

So the reason that surface roughening does not help give extra adhesion via extra surface area is that there is no (significant) extra surface area. Similarly, it cannot create mechanical interlocking. If I take the (still exaggerated) graph at 125 µm and invert part of it, clearly those surfaces aren't going to be locked together (Figure 3.10).

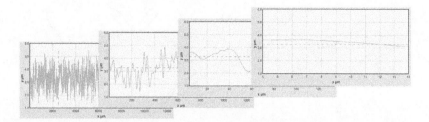

Figure 3.9 As you expand the scale in the *x*-direction you see that the roughness is a myth – it is actually a remarkably smooth, gently sloping surface.

Given that roughening a surface does not help adhesion via extra surface area or interlocking, why is surface roughening so often recommended? Partly out of habit from the myth, partly because a light roughening causes no harm and because many surfaces contain a lot of unstable junk which isn't very well adhered. The junk might be oil, dirt, surface damaged by UV light or maybe some weird hydrated oxides on metals. Rubbing the surface to return it to a clean, solid condition allows the normal surface energy processes to work.

There is one clear exception to my statement that mechanical interlocking is useless. If you apply an adhesive to paper, cardboard, cloth or wood, and if you allow the adhesive to flow the necessary multi-μm lengths to wrap around the fibres, and if the adhesive has significant mechanical strength, then you really

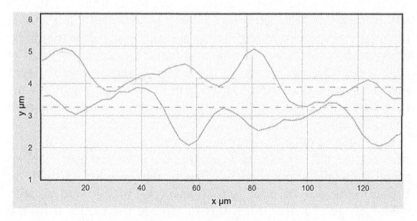

Figure 3.10 These two rough surfaces cannot mechanically interlock.

Figure 3.11   Mechanical interlocking when the adhesive can flow into and
wrap around fibres.

can get mechanical interlocking (Figure 3.11). Adhesion science
for these systems is much more about getting the flow and
mechanical properties right than it is about how things, in
general, stick to each other.

3.7   SOLDERING

Another form of liquid glue is molten solder. If you solder badly
(as I have done rather too many times), the result is a pure
surface adhesion bond that can look OK and which gives
adequate strength if you give it a vertical pull, but which fails
when there is a peel force. Good soldering is more than surface
adhesion. First, it requires the use of a "flux" that dissolves
and/or "reduces" the oxide contaminants on the surface to produce the native metal. Then the metals in the solder need to join
up with the metal (typically copper) to create a metallic bond, i.e.
where electrons are shared across all the metal atoms. This will
still not provide a tough joint because the electrons are shared
unequally across dissimilar metals and there is a clear boundary
where stresses can concentrate and a crack can propagate. The
trick is that the molten tin in the solder is able to dissolve (not
melt!) a few of the surface atoms from the copper. This means
that the boundary is tin/tin-copper/copper which, by not being a
sharp boundary, is more able to cope with crack stresses.

  Brazing is somewhat similar. Two high melting point components (usually steel) are separated by a narrow gap. Molten
metal (typically copper) with a lower melting point is made to
flow into the gap. On cooling there is a solid metal bond. As with
solder, there is a need for a flux to remove the oxides from the
metal surfaces. The use of lower melting point metals means

that there is less thermal strain on the components, and the brazed area can be larger than a welded joint, leading to a more distributed set of stresses. The bonding metal is of higher melting point and modulus than the necessarily more gentle solder, resulting in a bond that is likely to be more durable than soldering.

3.8 WHAT HAVE WE LEARNED SO FAR?

If you are a gecko, surface energy adhesion is ideal. If you want repositionable hooks on your wall, surface energy adhesion can sometimes work. If you want to fix something at home and you aren't going to put unexpected peel loads onto it and if you can find a liquid glue that rapidly changes to a solid, then surface energy adhesion works OK.

If we recall that surface energies change, across most relevant materials, only by a factor of 2.5 across most relevant materials, then we cannot get a 10× increase by increasing the surface energy. This means that we need some very different science to help us. We will find that there are two very different approaches to strong adhesion, one of which relies on having weak adhesives!

Before we start exploring strong adhesion, we need to understand how to determine whether adhesion is strong or not. We already have a hint that this is not an easy matter. The two strong men found that the adhesion was strong, then the little girl showed that it wasn't.

In most adhesion science books, the question of measuring adhesion is kept until later, because it is assumed that measurement is something rather straightforward. My view is the other way around. We cannot understand what makes strong adhesion without understanding that there is no objective way to measure it! Very quickly we will find that the force needed to separate the same samples of pure rubber, held together by surface energy, can vary by more than 1000× depending on how the force is applied.

CHAPTER 4

How Stuck Is Stuck?

It seems very easy to know how well two things are stuck together. You measure the area of overlap then measure the force needed to pull them apart. Force divided by area in standard units is $N m^{-2}$, also known as a Pascal, Pa. Because a Newton is rather small and a square metre is rather large, a Pa is small and adhesion values tend to be quoted in MPa or GPa.

Let us start with the system we introduced in the chapter above, with gecko-style, glueless, surface energy adhesion. Take two pieces of smooth rubber (rubber strips can be cast onto glass to produce super-smooth surfaces) and place them together so they spontaneously stick to each other.

We can pull them apart in three ways (Figure 4.1):

- Pull up one end and peel the samples apart: Peel test
- Pull along the join and try to shear them apart: Lap shear test
- Attach something strong to the back of one piece and pull it up vertically while holding the lower piece in place: Butt test

Rubber typically has a surface energy around 40 mJ m^{-2} which, via a bit of manipulation, can be seen to be equivalent to a peel force of 40 mN m^{-1}. If we have a sample width of 25 mm (0.025 m) then we find that the peel force is $40 \times 0.025 = 1$ mN. With an overlap length of 25 mm, the area is $0.025^2 = 0.000625$ m^2 so the

Sticking Together: The Science of Adhesion
By Steven Abbott
© Steven Abbott 2020
Published by the Royal Society of Chemistry, www.rsc.org

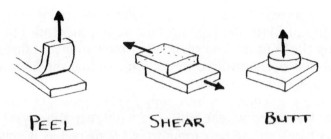

PEEL   SHEAR   BUTT

Figure 4.1 The three types of adhesion tests: Peel, Shear and Butt. Even for the same material the force need to separate them differs by factors of 100s or 1000s.

adhesion value in $Nm^{-2} = 0.001/0.000625 = 1.6$ Nm^{-2}, a very small adhesion value. Readers may notice that including the overlap length is a cheat; it is the only way to get a value in the same units as the others. This is an early indication of the problems ahead. I am cheating deliberately and openly; many in the adhesion world are unaware that these numbers are meaningless.

If you measure the force required to fail the same joint by pulling in shear mode, you find it needs a force that depends not only on the surface energy but on the thickness and the modulus of the rubber, i.e. how much stretch you get divided by the force required to create that stretch. Using typical values, we find that $F = 1.25$ N (not the mN of peel!) giving a $Nm^{-2} = 2000$.

If you now try to pull it apart vertically, in the butt joint test mode, it requires $F = 27$ N and a $Nm^{-2} = 55\,000$, 30 000 times more than the peel. This now explains why the two strong men could not pull two pieces of rubber apart (butt joint) while the little girl had no problem (peel joint). Although the men were stronger, they weren't 30 000 times stronger!

You can find all the formulae (which come from Prof. Kendall) and play with the key values in an app I wrote: https://www.stevenabbott.co.uk/practical-adhesion/weak-strong.php. More important is the message that Adhesion is a Property of the System. If your system is one where the forces are always equivalent to a butt joint, then the 55 000 Nm^{-2} might be good enough. Yet if there is any risk of some peel forces, then the 1.6 Nm^{-2} is likely to be very worrying.

Why do the values change by more than 30 000? Because joints fail whenever the local forces exceed a limit. In the peel test, the forces are focussed exactly at one point – the point where the joint will fail. With the lap and butt joints, the work put in to breaking the joint is also expended in stretching and distorting the material layers involved in the joint so that the force at the edge of the joint is much smaller.

Still, these calculations were only for two pieces of rubber in contact. What about real adhesion with real adhesives? This takes us nicely to the next topic which is as real as it gets. But before we go on, it is useful to see some real test equipment in operation.

https://youtu.be/8Gu0lbwbybw In the video, taken in Sugru's labs, we compare a shear and peel test on the same material which is a silicone adhesive well-adhered to stainless steel. In this example the shear was (for the same area) 2.3 times stronger than the peel. Why only 2.3×? Because although the test is labelled "peel", the forces involved are rather mixed so it isn't a pure peel test.

4.1 KEEPING THE WINGS ON THE PLANE

When we take off in an aeroplane, we rather expect that the wings will stay in one piece. Although the wings are bolted (not stuck) onto the fuselage, the complex wing assemblies are largely held together by adhesives. The manufacturers need to know if the adhesives are strong enough. This means that engineers run whole batteries of tests to see what the limits are. They then figure out how to specify and validate products to ensure that they reliably meet the practical requirements with a good margin of safety.

It would be rather wasteful if each aircraft manufacturer had to make their own estimate of how good the adhesive should be. In practice the manufacturers (and regulatory authorities) worked together and created an industry standard test, saying that any adhesive that withstood a specific load under specific conditions would be good enough. As far as I know, no wings have ever fallen apart with glues that passed this test. What, then, is this test?

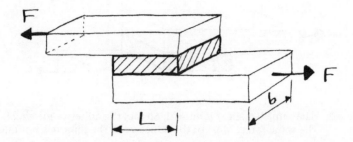

Figure 4.2 The lap shear joint test. Despite its name, it doesn't depend much on the overlap, *L* and doesn't fail in shear.

The main forces acting on the wing assemblies are shear – across the joint rather than up from the joint which is peel. So, the standard test is the lap shear test (Figure 4.2).

Two pieces of aluminium (up until recently, the standard material used for aircraft wings) of width *b* are overlapped by length *L* and the space in between is filled with the adhesive. Once the adhesive has set, a force *F* is applied to each end and increased until the joint fails. The contact area is *L*×*b*, so the resulting test value is said to be (soon you will see why this is wrong) $F/(L\times b)$, in $\mathrm{N\,m^{-2}}$.

To almost everyone that is the end of the story. It is obvious that the longer *L* is, the stronger the bond (there's more adhesive doing more sticking!) and equally obvious that the bond will fail in shear, because that's what the test is doing.

https://youtu.be/UhAJsq-Xmb8

Both "obvious" statements are wrong. Above a certain minimum value, *L* makes no difference. The video shows this. The force needed to fail with a lap of 4 cm was not much smaller than an 8 cm overlap. When we went down to 1.5 cm the force *was* smaller, but was still half of the original, not 20%. Although not clear from the video it is also found that lap shear joints fail in peel.

One of the more delightful academic papers on lap shear joints compared the failure forces *F* for joints with the same *L* but with more and more of the adhesive removed, replaced by inert plastic discs of the same thickness. They were able to

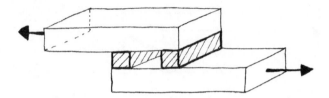

Figure 4.3 The same lap shear joint with 60% of the adhesive missing fails at the same force. Most of the adhesive in the joint has no function!

remove two thirds of the adhesive before there was any reduction in the failure force – though only if they removed it from the centre of the joint (Figure 4.3).

The failure in peel mode is of interest to those who have flown in a Boeing 787 or Airbus 350 where the wings are made not from aluminium but carbon fibre. Next time you are on one of them you can amaze your fellow passengers by telling them, correctly: "This plane fails an industry standard test for adhesion of its wings". If they are interested, you can provide the following explanation.

If you take a section of a carbon fibre wing and subject it to the same test as the one used to validate all previous Boeing and Airbus planes you will find that it fails with a much lower force, and a visual inspection shows that the whole structure has peeled apart rather than shearing along the joint. Carbon fibre is used in modern aircraft because it is astonishingly strong along its length – and that's what matters. The fibres resist strong pulls. Across the sample is a different matter – it can be peeled apart relatively easily because the fibre-to-fibre strength is not especially high; it doesn't need to be. Why, then, are we allowed to fly in planes that fail an industry standard test?

When the aircraft lap shear test was developed, the experts fully understood that it was really a test that failed in peel. Even the aluminium samples fail more in peel than in shear. But it didn't matter, because the aluminium itself does not peel apart, just the joint. The test is very easy to set up and as long as they agreed on a failure force high enough, the fact that it wasn't measuring true shear failure was of no importance. When the carbon fibre planes came along, the experts were aware that they would fail the "standard" test so they skipped that and went straight to the less convenient, but pure shear, "double lap test"

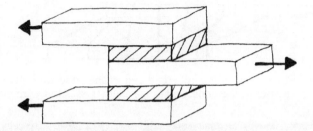

Figure 4.4 The double lap shear test. This *does* fail in shear, but the amount of overlap is still largely irrelevant.

(Figure 4.4). In pure shear, the great strength of the carbon fibres along the shear direction is a big advantage and the wing adhesives pass the test with no problem. So, yes, you are OK, the wings of the 787 and A350 are safe to fly with.

4.2   THE EXPLANATIONS

We have two things to explain. First, why most of the adhesive in a lap joint does nothing very much. Second, why the shear joint fails in peel.

The first effect is easy to explain.

With no loads on the joint, the adhesive can be considered to be like a mat, or a patch, sitting at rest within the overlap between the two adherends. A force is applied on both the upper and the lower adherends: the adhesive mat prevents them from flying away, and it begins to tip and shear as the adherends pull in opposing directions: the lowest portion of the adhesive follows the lower adherend and the upper portion of the adhesive follows the upper adherend. What also happens is that, thanks to the locking action from the adhesive, each of the ends of the adherends within the overlap stay more or less where they are, while the other ends stretch further and further away from the joint as the load increases: they stretch at a rate determined by their modulus. The edge of adhesive nearest to the free end of each adherend sees a significantly higher stress than the bulk of adhesive near the overlap. Crucially there is a substantial section in the middle of the joint where there is little force (Figure 4.5). A calculation shows that this effect is very pronounced, with the stresses highly concentrated near the ends (Figure 4.6).

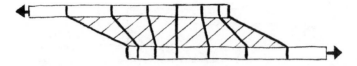

Figure 4.5 The stretch in the adherend and in the adhesive. There are large stresses at the ends, and only small stresses in the middle.

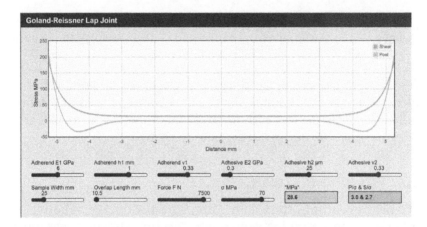

Figure 4.6 A model of the lap shear joint shows that stresses are nearly absent along most of the joint, and highly concentrated at the ends. The peel stresses overtake shear right at the end.

Because the concentration of the forces near the end is due to the stretching of the adherend, the less it stretches, the less the concentration and, therefore, the stronger the joint. The strength of a joint, yet again, depends as much on the adherend as on the adhesive. The very high modulus of the carbon fibre components means comparatively little stretching, and therefore a strong joint.

The second effect can readily be observed with a simple lap joint. There is a strong twisting motion at the joint ends; this sends forces in the peel direction (Figure 4.7). When you do the

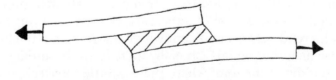

Figure 4.7 It is the distortion of the whole joint that creates the peel forces which cause the lap "shear" joint to fail.

calculations, in most cases the stress right at the end is higher in the peel than in the shear direction.

There is one more simple corollary from this: When someone says that a lap shear joint showed a failure *stress* of X $\mathrm{N\,m}^{-2}$, their claim is meaningless. They have found that a force F over a lap area of A was sufficient to cause failure. What they cannot say is that the failure stress is F/A. If they doubled A by doubling L, F might increase by, say, 10%, so now the "failure stress" is 55% of the previous value. The absence of a direct link between F and A means that F/A values are meaningless. In the tests where the adhesive was removed from the middle, the F/A values are, again, meaningless.

If you are sufficiently interested to explore the full lap shear app (https://www.stevenabbott.co.uk/practical-adhesion/g-rlap. php), based on the theory of Goland and Reisner, you will find (and it should be no surprise) that the stress concentrations do not depend just on the strength (modulus) of the adhesive. The thickness of the adhesive and the thickness and modulus of the adherend also affect the stress concentrations and, therefore, the chance of failure. Someone interested in strong adhesion in a lap joint might find that changing the thickness of the adherend was as important as changing the adhesive. This again tells us that F/A values are meaningless because they in no way capture the full complexity of the test. If F/A changes (and it does!) when the adherend thickness is changed, that tells you nothing useful about the adhesive itself.

As you are starting to see, Adhesion is a Property of the System is not some cheap slogan – it really matters. If you start believing "failure stress" values of lap joints, you will start believing that you should not fly in a Boeing 787.

4.3 NEVER RELY ON A BUTT JOINT

Adhesives manufacturers advise us to use the adhesive "sparingly". The reason is that for a butt joint, where we are pulling up vertically, the force F required to break the joint depends on $1/\sqrt{d}$, where d is the thickness of the adhesive layer (Figure 4.8). (The full formula is shown and explored in the app: https://www. stevenabbott.co.uk/practical-adhesion/butt-test.php.) This tells us that the adhesive is acting as the weakest link and that the

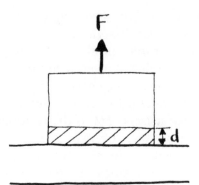

Figure 4.8   In a butt joint, the thinner the glue layer (smaller *d*) the stronger
the bond. That's why we are told to use the glue sparingly.

ideal is the lowest-possible thickness – though there is a limit
beneath which the assumptions behind the formula break down.

As before, you will find people quoting F/A as $N\,m^{-2}$ to describe
the butt failure stress and, as before, this is meaningless. The
stresses when you pull on a butt joint are focussed very sharply
around the edges and once there is failure at the edge, the whole
joint fails (Figure 4.9). This is why you should never rely on a butt
joint. Think about it. Suppose there is a little bit of dirt in the
middle of the joint. That's not a problem. That same bit of dirt
just at the edge means a defect in the adhesive at that point – the
weakest link – just where the stresses will be at their highest.

It is even worse than that. The theory assumes that the pull is
exactly vertical. The real world is not as kind as theoretical joints.

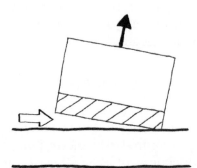

Figure 4.9   Butt joints are unreliable because the slightest tilt of the force can
focus the crack energy on one portion of the joint, causing it to fail
early.

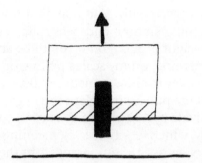

Figure 4.10 A dowel rod within the butt joint can help keep the pull forces vertical, helping to avoid over-stressing one part of the joint.

The pull might be a few degrees away from the vertical. Or the force might not be applied absolutely centrally. This means that the stresses are concentrated even more strongly at one portion of the edge, so failure is even more likely.

This is why many of our home repair jobs fail. It isn't just because we chose the wrong adhesive or didn't squeeze enough or didn't leave it long enough before testing. It might equally be because our fix was a butt joint (e.g. a knob on a difficult drawer) and a slight bit of dirt or a slight mis-pull is enough to wreck the whole joint.

It is quite common to add a piece of doweling into such a joint as in Figure 4.10. There may be some adhesive around the doweling and that might help a little, but the main reason is to stop any tipping; this makes the joint much stronger in practice.

4.4 AT LEAST OUR ADHESIVE TAPE IS NICE AND SIMPLE...

Peel a piece of household adhesive tape from its handy roll and stick it to something convenient. Done! Nothing to it. It is just a piece of sticky stuff, what else do we need to know?

In the chapter on these sorts of "pressure sensitive adhesives" (warning, despite their name they are not pressure sensitive) we will find that the way they stick is far from simple. Here we will look at the apparently simple task of measuring how sticky they are.

If you have some digital kitchen scales, clean their surface then stick some tape onto it, rubbing it down gently to get rid of obvious air bubbles and leaving a free end on which to pull. Zero the scales then start pulling up vertically. You may find that you lift up the scales. Find something smooth and heavy such as a plate or dish so that the extra weight is enough for you to be able to pull up on the tape. After re-zeroing, on my scales pulling on an 18 mm wide tape caused the display to show something like -360 g, the minus sign meaning that the pull is equivalent to a negative weight. To go from g to N you divide by 100 (divide by 1000 to convert to kg then multiply by gravity, which is ~ 10 m s^{-2}), meaning in this case that my force was 3.6 N across 18 mm. Across 1 m the force would be $3.6 \times 1000/18 \approx 200$ N m^{-1}, a normal value for such a tape. What you have done is a crude form of a standard 90° peel test. Proper testers use tricks with sliding sample holders to ensure that the pull is always 90°, and other machines do the test by peeling at 180°. The video taken in the Sugru labs shows some typical professional test equipment. Whatever the equipment, they will be doing the same thing: measuring the force needed to peel the tape, probably taking an average to even out the wobbles in the measurements, and ignoring spikes at the start and finish. Our kitchen scales version is good enough for our purposes.

https://youtu.be/d3iSrBcQPjg and
https://youtu.be/hTwnZo0IJo0

Now put a fresh piece of tape down and place a block of something from the freezer on top of it to cool the adhesive and re-test moments after removing the cold item. Similarly, place a fresh piece of tape and, using your judgement to avoid overheating the tape or your scales, heat the tape with a hair dryer or by putting something like a saucepan with warm water on top, and re-test. Or, as in the video, use a freezer and oven respectively. Depending on the tape and the temperatures you reached you will find that the peel value can change significantly. In my own tests for the video I was shocked to find that the specific tape and temperatures did not give me the lower adhesion I had expected at both low and high temperatures. I clearly chose the wrong tape for producing a good demo! There are, I assure you, plenty of tapes that would have failed at these temperatures.

The adhesion of any tape depends on the temperature (it will fail at low enough and high enough temperatures). That's not so much of a surprise. Now we need to repeat the peel tests at faster and slower speeds. This is easy to do on the real machines (as shown in the Sugru video) and harder with the kitchen scales. Those of us who have had a plaster stuck over a wound already know what happens. If the peel is slow, the tape sticks painfully well. If we have the courage to do so, or if a nurse does it for us, a very quick removal is almost painless because the adhesion is much less. If you do this scientifically you can find that the peel force at a faster speed is reduced by the same amount as a test done at a lower temperature.

https://youtu.be/7yfWbzpRc_M

The first few takes for the video of one of the "animal tough" strong tape were abandoned because the tape kept skipping from a strong slow peel to a rapid near-zero peel. Eventually I realised that it was important to capture this effect, while still showing the significantly stronger peel. I could then repeat the setup with a deliberately fast peel – and the tape came off instantly.

From all this we work out that "the" adhesion of a simple piece of adhesive tape depends on the temperature and on the speed. The interdependence of temperature and time is a fundamental principle of many systems and comes with acronyms such as TTS (Temperature Time Superposition), TTE (Temperature Time Equivalence) and WLF (Williams, Landel and Ferry, who found a general formula for the effect). Interested readers can explore more about WLF via an app: https://www.stevenabbott.co.uk/practical-adhesion/wlf.php.

In the PSA chapter, Chapter 6 we will find that rough surfaces generally reduce the peel strength, contrary to the usual idea that roughening a surface can help adhesion.

There is one more thing to discuss about peel in this chapter. For a given adhesive, peel depends strongly on the thickness and modulus of the backing tape to which it is stuck.

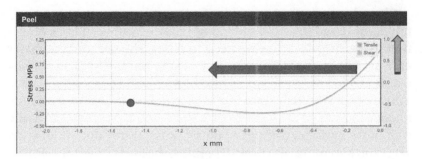

Figure 4.11   Peel seems simple until you model it. The forces can be spread far
ahead of the peel zone (this helps increase the effective strength)
and they even produce a compression zone ahead of the zone
experiencing the peel tension.

You can calculate the forces just ahead of the peel zone of an
adhesive tape (those interested can find the app at www.
stevenabbott.co.uk/practical-adhesion/peel.php) and it turns
out that they are spread over more than 1 mm (in Figure 4.11
they reach zero at ~1.5 mm).

One of the fascinating aspects of the zone ahead of the peel is
that part of it (after ~0.3 mm in the screen shot) is under compression, i.e. pulling up on one part of the tape causes another
part to be squeezed downwards. This compression zone is important for understanding silicone release – but you must wait
for the PSA chapter, Chapter 6 to know why.

Even more interesting is that the forces around the peel zone
don't just depend on the properties of the adhesive. Not only the
thickness of the adhesive but also the thickness and modulus
of the backing tape have an influence on how the forces are concentrated. If the backing tape is a relatively weak polyethylene or
polypropylene (compared to the more usual polyester) and/or the
backing tape is thinner, then the stresses are more concentrated so
the adhesive will fail more easily. Taken together, this means that
our peel force depends on temperature, speed *and* on the backing
tape material *and* on its thickness. Yet again we cannot say what is
the true adhesion of an apparently simply piece of tape because the
answer is that Adhesion is a Property of the System.

The world of PSA has its own range of tests which will be discussed in the PSA chapter. Here I mention just one, the delightful
Rolling Ball test which lets a ball pick up speed by rolling down a

standard incline before running along a horizontal strip of tape. The stickier the tape, the shorter the distance the ball travels before stopping. The test is completely unscientific, in that there is no way to relate the stopping distance to any rational physical parameters of the adhesive. Like so much in adhesion testing, it is used as a convenient *relative* test. If the ball has always stopped at 20 cm for tapes made over the past month yet today it stops after 15 or 25 cm you don't know *what* has changed about the tape, you just know that it is different and therefore you have a problem in production.

4.5 THE TAPE TEST

Returning to more conventional adhesion, we can use adhesive tapes as the key part of the least-scientific and most-used adhesion test of all. It is one you can carry out yourself. Let's say you want to test whether a paint is strongly adhered to some substrate. You wait until the paint has dried, apply a strip of tape to the surface of the paint, and pull. If the tape pulls off some of the paint, then you know you have a problem.

Imagine that you were the unscrupulous sales rep of this paint and wanted to demonstrate how wonderfully it stuck, even though its adhesion was known by you to be poor. Here are some tricks. Sadly, I know that each of them has been used, though never, I assure you, by myself:

- You could use a really weak tape and/or place it down very badly
- You could get your formulators to add a bit of silicone oil to the paint so that it came to the surface and gave very poor adhesion to the tape
- You could make the paint rough ("a special matt finish look") so that the tape could not make good contact
- You could pull very fast, reducing the adhesion in the same way that a nurse reduces the adhesion of a plaster
- You could apply your test to the middle of a large area of paint, away from any cracks or edges

That last trick arises because in practice the tape test, as far as the paint is concerned, is a butt test if there are no nearby edges, and butt tests without stress concentrating edges are easy to pass.

Those who want to carry out a fair test use an industry standard tape, designed for such tests, on a cleaned surface (no oils to fool them), not too rough, with a controlled pull speed designed to maximise the tape's adhesion and, crucially, on a portion of the paint that has been scored with some knife blades to guarantee some pre-cracks. Although this version of the test is still very crude, it is often the *only* viable test. Years of industry experience has shown that coatings which survive a fair tape test are unlikely to fail in the field.

You might wonder about the phrase "guarantee some pre-cracks". Let me explain with two specific examples.

1. I needed to cook some fish that came in a nice pack with an "easy peel" film on top. These packs come with one corner pre-cracked so you can grasp the film and pull it apart. For this pack they had failed to provide the crack. I had to attack the pack with various scissors and knives to be able to get the slightest crack required for me to peel the film. It was astonishingly difficult. When I eventually managed, the film peeled off easily – yet without the starting crack, the adhesion was perfect.

2. As shown in the video, a piece of super-strong tape applied to the back of a roll of ordinary tape cannot peel it apart – unless it is applied around the "crack" in the tape that marks the fresh end of the tape. We all know how hard it can be to pick at the end of a roll of tape to get it started. A piece of tape stuck onto the end is a good way to do the job.

https://youtu.be/8w8TgP1Kz40

An extreme example of adding a pre-crack before a test comes from the auto industry. Brake pipes are safety critical yet readily corroded by salty water. They are protected with a layer of, say, zinc, which in turn is protected by a primer and a layer of polyamide (nylon). One test is to flat-grind the layers until the bare steel is reached, then to put that sample into a corrosion chamber. After many hours of salt spray, although the exposed steel will corrode, the primer and polyamide must not fall off. That is one of many examples of why modern cars last so long and remain so safe.

Many complex adhesion tests start with a known crack. For example, a wedge test requires a wedge to be hammered in to the joint before loads are applied to see how easily the crack opens. Analysis of the results of crack tests are difficult because it is hard to distinguish "crack" effects from "adhesion" effects.

4.6   BRING IN THE EXPERTS

Maybe this insistence that we cannot measure a true adhesion is because I haven't dug deep enough, or I haven't found the right equation. It could actually be my own fault.

Well, it's not my fault. My first evidence comes from Dr Robert Lacombe, a guru from IBM's research labs who wrote the standard work on Adhesion Measurement Methods. The first big chapter is, in his own words, a Consumer's Guide (for a UK audience the equivalent is Which Magazine) analysis of measurement techniques. For each technique (just as in the guides) all the good aspects of the technique are described; then there is a "but" and all the bad things are listed. In the guides you normally come to at least one product which is pretty good all round. In Lacombe's book, all the techniques come over as being full of problems.

My second piece of evidence came after I went to a lot of trouble to find papers from top research groups who might be able to show objectively that test A was better than test B. In each case the papers took a well-defined system and measured the adhesion with a variety of techniques in well-equipped labs, trying to find some objective ways to compare and contrast the various results. In all cases they failed. We have already noted that a $N\,m^{-2}$ value from a measurement is usually meaningless because the N's are not being applied equally over the m^2 of the test sample. These academic papers confirm that there is no meaningful way to go from the values of one test to those of another. Even worse, if they tested a range of samples prepared by different methods, the strongest adhesion judged by one test would usually have no relation to the strongest in another test.

To make things even more complicated, the next example shows that sometimes an adhesion failure has nothing to do with a failure of the adhesive.

4.7   BOIL IN THE BAG

In recent years, boil in the bag foods have become common. The food is in a plastic pouch which is put into boiling water until cooked. I was once asked to solve an adhesion problem in one of these systems. The bags are complex structures with various polymers, adhesives and a layer of aluminium to act as a barrier to oxygen which would spoil the food if it diffused in through the pouch. In this problem, the adhesive to the aluminium was failing in tests designed to simulate people tearing open the pouch once it has been boiled. It is fairly easy to come up with reasons why the adhesive might be failing in some pouch designs and not in others. In this case, however, it turned out that "good" pouches differed from "bad" pouches not in terms of the adhesive and layers next to the aluminium, but in the type of polyethylene used a few layers over. One type, "HDPE", gave good tear opening, the other type "LDPE" gave delamination at the aluminium interface. HD (High Density) polyethylene is more crystalline and stronger than LD (Low Density). When you tried to tear open the pack made with the HDPE, the whole area remained relatively rigid so the tearing forces went where they were supposed to go, and the pack opened nicely. With the weaker LDPE, the tearing forces distorted the pack, sending some forces along the interface with the aluminium, causing it to fail.

If we had asked the question "Is the adhesion to the aluminium good enough?" we might have spent a lot of time doing classic adhesion science tests and found that the answer was "Yes". Instead, we spent more time thinking of the system, where the answer turned out to be "Yes, good enough for HDPE but not for LDPE". Adhesion is a Property of the System.

4.8   TESTING AT HOME

In Chapter 1 we discussed the difficulties we have in knowing whether our adhesive will be good enough for the job. We cannot do repeated scientific tests on, say, an orchid-watering jug. But even if we had a lot of test equipment and a lot of broken jugs, we could spend a lot of effort and still not reach any good conclusion. If we did lots of butt tests, yet the sample was mostly exposed to peel, those tests would be worthless.

The fact that most adhesion tests are useless is good news. It means that we can use our understanding of adhesion to make smart choices rather than trying to find a suitable test. I once broke the handle of a friend's mug. It happened to be a mug from a museum with some family historical connections. The scientifically best way to fix it was to organise a day out with the friend to be able to buy a replacement mug from the original museum shop. That's an extreme example, but the point is a real one – to think of the science of the whole system (rather than run to the adhesives cupboard) to find the best way to solve the problem.

The next three chapters give you the science you need to make your own smart choices for whatever it is you are trying to stick together.

CHAPTER 5

Strong Adhesion

It seems obvious that one good way to get strong adhesion is to line up a bunch of strong chemical bonds across the interface. This, like so many intuitions about adhesion, turns out to be profoundly wrong – it is *not* a great way to get robust adhesion.

To put it another way, too much of a good thing is a bad thing. In this case, too many chemical bonds are not what you want for strong adhesion. We will find that 1% of chemical bonds across an interface is closer to being optimal than 100%. If this was not the case, we would be in big trouble because many of the strong adhesive systems we rely on stand no chance of gaining a high percentage of bonds across the interface. The fact that they naturally create just a few per cent of bonds is why they work so astonishingly well.

It seems, then, rather important to understand why having a lot of chemical bonds is bad for adhesion. Fortunately, we can readily understand the problem by (in our imagination) throwing a brick at a window.

Let us pretend that we have joined two pieces of glass together by creating a perfect row of silica bonds between the two pieces – creating what looks like a perfect window. In reality, we have just taken a normal piece of glass, but it is valid to imagine that we had created such a perfect line of such bonds. If we try to pull the glass apart, we find that it is very strong. Now with a sharp diamond, scratch along our perfect line of bonds. A few of the bonds are

Sticking Together: The Science of Adhesion
By Steven Abbott
© Steven Abbott 2020
Published by the Royal Society of Chemistry, www.rsc.org

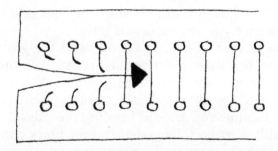

Figure 5.1 Once you break one bond at a crack at a rigid interface, there is nothing to stop all the other bonds breaking.

broken by the point of the diamond, leaving all the others alone. Now slightly bend the glass at that point and, crack!, our entire adhesive bond is broken (Figure 5.1). Our brick through the window does the same thing. One crack and the whole window disappears.

If you were able to measure the energy required to break the glass you would find that it was $\sim$1 J m^{-2}, equivalent to a peel force of 1 N m^{-1}. If you recall the experiment with the adhesive tape on the kitchen table, we had a peel force of 200 N m^{-1}. A trivial piece of adhesive tape, with no chemical bonds, is 200$\times$ stronger than the glass interface which was created from pure chemical bonds. The 1 J m^{-2} is a number described in a famous quote from Gordon in his book, *Structures: or why things don't fall down*: "*It turns out that the total energy needed to break all the bonds to any one plane or cross-section in most technologically relevant materials is very much the same and does not differ widely from 1 J m^{-2}*". Another quote from Gordon is relevant, especially when we remember that bullet-proof glass relies on a layer of a rather weak polymer in the middle: "*The worst sin in an engineering material is not lack of strength or lack of stiffness, desirable as these properties are, but lack of toughness, that is to say, lack of resistance to the propagation of cracks*". The relatively weak polymer provides the toughness that the strong glass does not possess.

We cannot create strong adhesion by brute force. The first epoxy glues were not especially strong, so chemists strove to make them stronger and stronger, successfully, right up to a point at which the glues became useless. They hit the same problem as with glass: the stronger glues became too brittle.

Later in this chapter we will see how they got past this brittle limit with a trick you can't apply to glass.

What does "brittle" mean? As you bend the glass near the point where the diamond created a mark, all the bending energy gets concentrated at a single point. This is more than enough to break the chemical bonds and therefore to break the glass at that point. You are still bending the glass, so the crack energy is still there, still focussed, and the crack continues – it moves through the glass at the speed of sound. There is nothing to stop it.

To find out how to get strong adhesion we need to take some small steps, starting with the word "intermingling".

5.1   INTERMINGLING

For the next few sections we will be concerned only with adhesion between two polymer surfaces – sticking two pieces of plastic together, something that is done on a massive scale in food packaging, automobile manufacture, toy making and a vast array of other applications. To understand how we can get off to a good start with a bit of intermingling we can go back to the kitchen scales to find out what happens without it.

https://youtu.be/dxD9cXtI2nc

Place a piece of cling film (or whatever you choose to call it) onto the digital kitchen scales and tape it in place with strips of adhesive tape. Now rub another piece of cling film into contact, keeping one end clear to be able to pull. Pull up and measure the (negative) weight recorded by the scales. I used a top strip that was 80 mm wide and I measured a 6 g pull-off force. This is 0.06 N across 0.08 m which means $0.75\,\mathrm{N\,m^{-1}}$. This is $16\times$ larger than I expected, but it is still very small.

https://youtu.be/UqVSSa3SOvI

Now, in an experiment worthy of a Nobel prize winner, find a pair of toothbrushes. Put one toothbrush on the scales with its bristles facing up. The one I used weighed 23 g. Now press the bristles of the other brush fully into the one on the scales. When I did this, I could easily lift the 23 g brush off the scales. Repeat the experiment but only push the bristles half-way in. This is rather hard to control, but when I tried it, I could only lift the lower brush up by around 9 g before they fell apart. Finally, and ideally as a thought-experiment, randomly remove half of the bristles from each brush before putting them fully together. You can make a reasonable guess that you will only be able to lift half the weight.

We can make a formula for our bristling-with-science toothbrush adhesive system. To give it a scientific touch, we will give the Greek letter Σ to one of the parameters which is the number of bristles per unit area of brush. We will use the letter L for the bristle-to-bristle overlap length which goes from zero when there is no contact to the full depth of the bristles if we press hard. And we will say that U is the energy required per unit length of bristle-to-bristle contact to slide past each other. Our work of adhesion, W, can now be given: $W = \Sigma \times L \times U$.

This is the famous de Gennes formula for "intermingled" adhesion and de Gennes did indeed get the Nobel prize for this and other formulae. You might think that adhesion cannot possibly be described by such a simple formula – and you are right. I once spoke to a famous chemist who knew de Gennes. "Ah, yes, de Gennes was a genius at finding deeply insightful approximate formulae which are right in general and wrong in detail. He was never interested in the details, he left that to generations of researchers who followed up on his ideas". Although the de Gennes formula misses out on many subtle details, overall it turns out to be a reliable guide to intermingled adhesion.

5.2 UNCOOKED SPAGHETTI

https://youtu.be/UgjYB3kR6q0

We can gain a quite profound understanding of intermingled adhesion and its limitations if we switch to uncooked spaghetti for our next round of experiments. As in the video, place some in a pot on the scales, then insert a similar-sized handful to different depths into the first bundle (our bristle-to-bristle interface length 'L' now becomes a pasta-to-pasta interface) and measure how the pull-out force changes with L. Equally, if you play with the bundle density, Σ, you will find that de Gennes was right. The adhesion is modest because intermingling only resists coming apart through friction. We use the specific word "intermingling" to distinguish this type of adhesion from the far stronger mechanism which requires us, soon, to cook the spaghetti.

Now, in your imagination, shrink the spaghetti down to the molecular size. We have bundles of rigid polymer sticking up from a surface. If we have another surface with similar bundles, then, where they meet, we can apply ΣLU and calculate the real work of adhesion. It turns out to be around 0.5 $\mathrm{J\,m}^{-2}$ for typical polymer chains of 1–2 nm length and a density of 1–5 chains per nm^2. This means that intermingling of polymers across boundaries gives us $10\times$ more adhesion than surface energy alone. Now we know what we need, how do we get nano-scale polymeric spaghetti to intermingle?

If you are sticking the same polymer to itself then a bit of heat or solvent is all you need. With the external energy loosening things up, the chains of one polymer start to accept chains of the other amongst them, to a lesser or greater extent. If the polymers are *very* dissimilar then they simply prefer to keep themselves to themselves and no amount of heating or solvents can persuade them to overlap by more than a fraction of a nm (though even this can be measured as a significant improvement over surface energy). This "diffusion" across boundaries to create adhesion is often dismissed by adhesion scientists who say things like: "Most polymers are not miscible so you cannot get significant adhesion in this way". They are certainly right about polymer miscibility. If you try carefully to see if two different polymers can be permanently mixed you will find that, in general, they prefer to separate given enough heat and time. Although this is true, it is also irrelevant. There is a formula from Helfand (see https://www.stevenabbott.co.uk/ practical-solubility/polymers-across-boundaries.php if you want to know more) which tells us by how much dissimilar polymers can overlap. This can be many nm, which is exactly what we need for modest adhesion. Even for very dissimilar polymers, the Helfand formula gives a fraction of a nm. All those adhesion scientists who have dismissed the "diffusion" intermingling mechanism have been, well, there's no other word for it, wrong.

One of the top adhesion experts, Prof. Richard Wool, who had the correct understanding about intermingling, did many careful measurements around the de Gennes formula. He also did some delightful experiments that were published in the only adhesion science paper I know of that acknowledges the help of an expert carpenter. Wool, his students (and the carpenter) took various planks of wood and nailed them together with different numbers of nails per unit area (Σ) and different lengths (L). They measured the friction of the nail to the wood, U. When they measured the forces needed to rip the planks apart and crunched the numbers, they found that nail work of adhesion was given by $W = \Sigma \times L \times U$. What I call "intermingled adhesion", Wool called "nail adhesion". The name is not as important as the two concepts associated with it:

1. This sort of adhesion is significantly stronger than surface energy
2. It is generally not strong enough for real-world adhesion, so we need something extra

What we need, instead of nails, is something equivalent to screws. Which gives us an excuse to cook some of the spaghetti we used at the start of this section.

5.3 COOKED SPAGHETTI

As in the video, cook your spaghetti till it is "al dente", i.e. soft enough to be eaten but not so overcooked that it is mushy and is stuck together with, essentially, starch glue, which doesn't currently interest us. Indeed, cook the spaghetti with a few drops of olive oil added to the water. This will end up coating each strand of the spaghetti, inhibiting direct spaghetti-to-spaghetti adhesion.

Put a pile of it into a pot on your scales then grab a fork and stick it into the pile. The fork can be taken as being the other surface of your adhesive joint. Pull up – and you will measure a large negative value – the spaghetti is so entangled that although the fork is in contact directly with just a few strands of spaghetti, those strands are pulling up other strands which pull up other strands. Your adhesive force is spread over the whole pile of spaghetti and is not focussed into small areas.

If you cook some more of the same spaghetti, but broken in half before going into the water, the entanglement is less strong, and you pick up less spaghetti with your fork. Shorter lengths give fewer tangles and, therefore, less ability to spread the forces across the pile.

Welcome to the world of strong adhesion. This principle of entanglement (which is different from intermingling) explains many otherwise confusing aspects of strong adhesion. I have already given a clue why this is the case by mentioning that the forces are spread over the pile of spaghetti rather than focussed. You may recall why 100% chemical bonds are so unhelpful: the forces remain focussed along the crack line, allowing the joint to break easily. With the spaghetti, and entangled adhesion in general, the crack energy is dissipated over a wide area, greatly decreasing the chances of the crack continuing along the joint.

With your fork do a few other tests. Try pulling up very slowly on the spaghetti. Oh, no, the strands are slipping past each other (if you have nice al dente spaghetti with the added olive oil). The negative load measured on the scales is near zero – we have low adhesion!

Now pull up very sharply. Very often the whole mass of spaghetti comes up at once, a very large negative load, the equivalent of strong adhesion (though be careful as it will soon all slide off and make a mess).

Does this mean that the measured adhesion depends on the *speed* of the adhesion test? Yes, it does! And we saw it before when we were peeling tape off our kitchen scales or imagined peeling a painful plaster off our skin. Speed-dependence is a universal property of adhesion science. There is even a number to describe such effects, named after Deborah, a prophetess in the Bible (see box).

Prof. Wool was especially expert at measuring the large increase of adhesion as he increased the polymer lengths, causing the system to change from intermingled to entangled. He could also confirm that the measured entangled adhesion increased strongly with speed of testing.

Deborah said (Judges 5:5) that *"The mountains flowed before the Lord"*. What she meant is that in the timescale of a major deity, a mountain is a liquid. What she didn't say is that if a mountain is hot enough it flows before people – i.e. solid rock becomes liquid lava at a high-enough temperature. What she didn't know (the experiments could only be done with great difficulty in the 20th century) is that if you measure water over a very short timescale it is as solid as steel. But she certainly knew that at a low enough temperature, water is a solid. Putting this all together we can define the Deborah Number, De, as The Natural Timescale of the Process divided by The Observation Time. So, if mountains flow over millions of years and your observation time is billions of years, De < 1 and the mountain is liquid. If water requires ns to move and we observe over ps timescales, then De > 1 and water is a solid. The natural timescales change with temperature, so De always needs to be specified at a given temperature. Although De is ≫ 1 for a mountain observed over a timescale of 1 hour, it is ≪ 1 when the mountain is at 1000° and you are trying to run away from the lava flow.

If entanglement is so important, it is a good idea to define it in scientific terms. As Wool pointed out, you could do all the calculations (and confirm them via experiments) using the simplest possible definition of "entangled". The diagram is my version (used with kind permission from Prof. Wool) of one that he created many years ago to show what the definition means. A system is entangled if a polymer chain crosses an imaginary line (which might be the real interface) three times. It is as simple as that.

In Figure 5.2, on the left we see that the chain crosses three times, marked with 1, 2, 3. If we pull up at the loop at the top, marked with the larger arrow, the bottom loop will clearly tangle (marked with the small arrow) with at least one other chain. Suppose the chain was cut short at the bottom right-hand loop, as in the image on the right. Now the chain crosses only two times and will easily be pulled out.

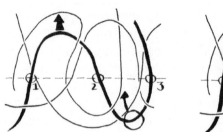 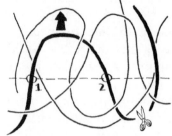

Figure 5.2 If a polymer chain crosses an interface three or more times it is tangled (left) because pulling up on the loop traps other chains. If it is merely intermingled (right), with just two crossings, the chain can be pulled out easily.

We can return to spaghetti. If it is uncooked or only lightly cooked, it is so stiff that it cannot get tangled. When it is cooked properly, it can easily tangle, but only if it is long enough. If you break your spaghetti in half or in quarters before cooking (or if you cut up a spaghetti portion for a child to eat more easily) it is too short to tangle.

Translating back to polymers, some of them just don't curl back on themselves very well so they are hard to tangle and need a very high molecular weight before they are significantly tangled. Others curl very easily and even a low molecular weight is enough to give entanglement. This is one reason why you can't use just any polymer as an adhesive. Those that tangle easily are generally better than those that find it hard to tangle.

To get strong adhesion, all you need is to get polymer chains entangled across the interface. The strong adhesion comes because when you try to separate the surfaces, the tangles spread the load across many nm, dissipating the crack energy as the polymer chains are dragged around. It is this dissipation that gives strong adhesion, not the "strength" of, say, chemical bonds, which, as we recall, on their own are not very effective. Where does the crack energy go? It heats up the polymer chains, harmlessly.

Most of us have created effective entangled adhesion one way or another. If you have ever heat-sealed a polythene bag or a laminating pouch, what you have done is melted the polymer at the interface to get the polymer chains to move across the interface and become thoroughly entangled, ready to give a strong bond on cooling. Anyone who has made a plastic modelling kit using a solvent-based glue has probably used one made from polystyrene. The solvent dissolves the surface of each part of the polystyrene model. When the two parts are put together, the polymers on each side are sufficiently mobile in the semi-solid state to tangle across the interface, leaving a strong, entangled bond once the solvent has evaporated.

Under the right circumstances, a coat of paint can entangle with the surface onto which it is painted. Similarly, printing of inks onto plastics often requires such entanglement.

What about the many cases where there is a different type of surface such as brick, plaster or metal? There seems to be no chance of entanglement. This is where things get really interesting. But first, we will stay with polymers for a little longer in order to understand how to make use of a different type of entanglement.

5.4 CROSSLINKS

So far, our polymer chains have been assumed to be like lengths of cooked spaghetti. For polymers like PE, PET and PC this is basically correct. For polymers like PMMA and polystyrene there are some side groups that create a sort of hairy spaghetti. That doesn't change the essence of these systems which is that each chain is free to move, within the constraints of whatever tangles are around it.

Many other polymers may contain long(ish) chains while, at the same time, having short links between chains. Such polymers, as described in Chapter 2, are said to be crosslinked. Epoxy glues are strongly crosslinked, household paints are crosslinked, rubbers are lightly crosslinked. We can look again at the image in Figure 5.3.

Depending on the ratio of crosslinks to long chains, these polymers will be more or less hard. Do not confuse "hard" with

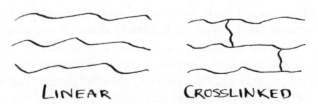

LINEAR   CROSSLINKED

Figure 5.3 In a linear polymer, the chains are independent. In a crosslinked polymer they are interconnected.

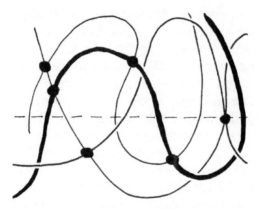

Figure 5.4   If the polymers merely cross at the points shown with a dot, the polymer is entangled. If the chains are chemically bonded at these points then the polymer is crosslinked. In terms of basic strong adhesion, it makes no difference whether they are tangles or crosslinks. Here we have a good balance of crosslinks and tangles. If every crossing was linked, the system would be too brittle.

"strong" or "tough". Many highly-crosslinked polymers shatter easily, while lightly-crosslinked rubbers can be very tough.

Let us now compare two systems (Figure 5.4): one where the crossings marked with a dot are ordinary polymer tangles and another where the dots represent chemical crosslinks.

If you start pulling the two systems apart across the interface, what differences do you find? The answer is that there isn't much difference. A polymer chain cannot simply pass through the other chain that is in its way, therefore making a chemical bond at that point does not add any extra degree of entanglement. In both cases, crack energy will be dissipated by the whole network moving.

At slow speeds there *is* a difference between entangled and crosslinked systems; with tangles the chains can gradually slip past each other. This means that physically entangled systems can *creep* over time, while crosslinked systems are more resistant. If we could crosslink spaghetti, we would, at slow fork extraction speeds, see a more negative value on our kitchen scales. The point is not that there is zero difference, but that the difference (at normal adhesion testing speeds) is not all that great. The fact that you have added some crosslinks does not automatically mean that the bond is stronger.

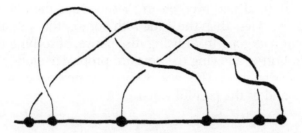

Figure 5.5 To tangle between a polymer and a non-polymer we need to create tangles via a few suitable chemical bonds. The rest of the polymer can tangle with the tangled polymer at the surface. The ratio of bonds to polymer is exaggerated in this image – a strong system will have much more polymer.

Now imagine if we have *lots* of crosslinks. This gives a rigid system, not really much different from glass, which can easily show brittle failure – the crosslinked system will be *worse* than the non-crosslinked system. Too much of a good thing is a bad thing.

5.5 STRONG ADHESION TO NON-POLYMERS

Because strong adhesion comes via dissipation through tangles, when sticking a polymer to a non-polymer such as metal, glass or ceramic, we need to find a way to create tangles between them.

We achieve this by creating a few bonds between the polymer and the metal so that the polymer is entangled to itself via the metal (Figure 5.5).

For simplicity, the diagram shows each polymer chain being tethered at both ends to give an impression of how tangled the system might be. In reality there would be far fewer chemical bonds and far longer polymer chains. This is important for two reasons:

1. It is very hard to get lots of polymer chains anchored to a metal
2. If you *could* get lots of anchor points, the adhesion would be very poor

Again, Prof. Wool has proved this second point. He made a polymer that did not stick very well to aluminium. He then added some special groups to the polymer which loved to form an anchoring point with the aluminium, as in Figure 5.5. As he added more of the special groups, the adhesion increased, as you would expect ... until at just over 1% of these groups the adhesion fell dramatically, to less than the value with none of the special groups. The system had gone from being dissipative, absorbing the crack energy, to brittle, allowing the crack to propagate along the interface. To see why this is the case, we need some elastic bands, as in Figure 5.6 and in the painful video.

https://youtu.be/CeyaUGZRbI8

Get hold of a bunch of rather weak elastic bands. Wearing a pair of gloves (when I first tried this, I didn't think this would be necessary), pull one band until you break it. Now loop two bands together and try again, then three, then four and so on. In each case, when the system breaks, only one of the bands has fractured. Yet if you have four bands, you have to stretch all four of them near to their limit before one of them breaks (Figure 5.6). This is hard to do (and without the gloves the snap-back is painful). Now apply these ideas to our adhesion to metal. If there are rather large loops of polymer between attachments, then you have to stretch a lot of loops before one of the polymer bonds breaks. It might be an attachment bond, it might be a bond in the middle; it makes no difference. If your polymer itself is crosslinked then you still have loops which can stretch and again

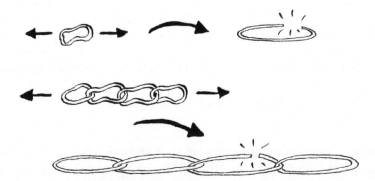

Figure 5.6   To break one elastic band we need to stretch it until it breaks. To break four of them, we need to stretch all four to the limit (which is a lot of work!) until one of them breaks.

just one of the bonds, a crosslink or not, has to break before there is failure. What we care about is that there are long segments that are free to stretch, absorbing the crack energy by causing other segments to move and dissipate the energy.

We can even provide a formula for this. The work of adhesion, W is given by $W = \Sigma \times L \times U$. Yes, the formula itself is identical to the de Gennes formula. Σ is the density of attachment points. L is the length of the loops. The difference is that U, rather than being an energy to overcome friction losses, is a chemical bond energy which happens to be $100\times$ larger than the friction values. This sort of entangled bond requires at least $100\times$ the work that can be stored in the system before failure. When you take into account all the extra dissipation from neighbouring loops, the work is more than $100\times$ larger, giving us strong adhesion.

The point of that formula is that if L is small, i.e. you have too many chemical bonds from over-crosslinking, then the adhesion is small, you are back to brittle failure. Too much of a good thing is a bad thing.

How do we get reactions between the metal and the polymer? Very often we use an adhesion promoter. A typical example is APTES which has an amino group (containing the nitrogen atom, N) at one end which loves to react with, say, epoxy or urethane adhesives, and a silane group (silicon, Si with three oxygens, O) at the other end which loves to react with the oxide surface of aluminium (Figure 5.7). Those who use APTES as a promoter for adhesion to aluminium tend to start by adding a small amount to the formulation, say, 0.25% and find a nice increase in adhesion.

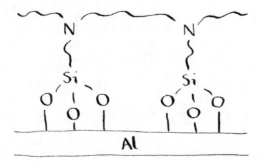

Figure 5.7 An adhesion promoter such as APTES can grab on to the aluminium via the siloxane bonds while the $-NH_2$ at the other end can integrate into an epoxy or urethane network.

They then try 0.5% and it is even better. When they go over, say, 1% they find that adhesion decreases drastically. This has puzzled many people (including myself before I understood the science). If more of something is a good thing, even more *must* be better! More is not always better. Too many adhesion promoting groups create a brittle interface.

When, in Chapter 7, we look at many of the adhesives you can buy in a store, the importance of "a silane group at the other end" will become apparent. Indeed, there is a whole class of newer adhesives that rely entirely on those silane groups to give the necessary strong adhesion.

Now we have the fundamentals in place for strong adhesion, we can look at some of the main classes of adhesives to find out more about their strengths and limitations.

5.6 CYANOACRYLATES – THE BASICS

We have already briefly discussed the cyanoacrylate superglues. They are liquids which polymerize when exposed to a little moisture contained in the air or on the surface of the things you are adhering. Specifically, the water creates a negatively charged hydroxide ion which reacts with one cyanoacrylate monomer, creating a cyanoacrylate ion which can react with the next and so forth (Figure 5.8). This rapid reaction solidifies the glue while in perfect contact with the substrates, forming an acceptable bond very quickly.

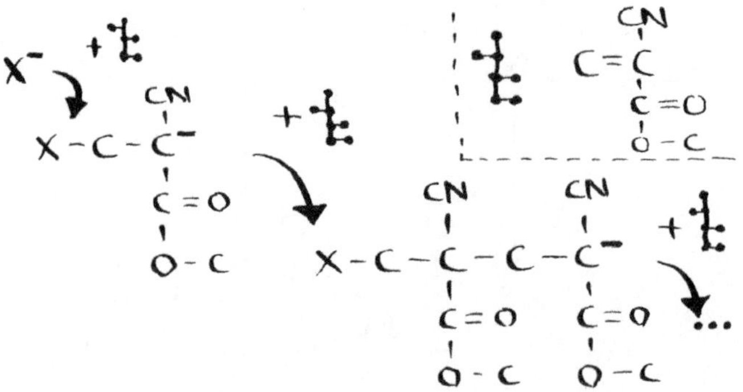

Figure 5.8   The X⁻ might be from some moisture or from an activator. Whichever it is, it starts the polymerization which just keeps going.

If you go to buy some superglue you will find a bewildering variety to choose from. How should you choose? Let me answer that with a little digression. I once spent 25 min (I timed myself) in a store trying to choose the best toothpaste. I am a scientist and I also happen to know some things about dental care so I decided that I would make an objective choice about which pack would provide optimal performance at a reasonable price. I carefully read the claims on each pack and also their ingredients. I happened to know that one ingredient, potassium nitrate, is specifically claimed to reduce the pain of sensitive teeth; as my teeth aren't sensitive, I didn't have to include that in my criteria. I also didn't want any aggressive whitener such as peroxide.

After 25 min I chose the same stuff I've always used because (a) it has a good flip top lid, (b) it comes in a size/price I find convenient and (c) the flavour isn't too bad. I made this choice because basically all the toothpastes are the same. Why, then, does each brand have a range of toothpastes that seem to be different yet are the same? It is all down to marketing. The marketers and the stores know that consumers *hate* having to choose between what are basically the same thing. Yet there is a sort of marketing war which dictates that you *must* provide a new, exciting product, even though it is neither new nor exciting.

After that digression you can guess my answer about superglue. Basically, all the superglues are the same and for the average fixing job you might as well use the cheapest – especially because you know that the tube will go solid before you have another job. After all, cyanoacrylates polymerize on exposure to moisture; opening and squeezing the tube allows at least a few water molecules back into the tube, kick starting the contents on their way towards full polymerization.

Scientifically, my answer is clearly false. There are real differences between grades. The first set of differences come from the cyanoacrylates themselves. You can get methyl (1), ethyl (2), butyl (4), octyl (8) ... cyanoacrylates where the numbers tell you the length of the chains of carbons attached to one part of the monomer. If you make the polymer from the octyl variant, it has more hydrocarbon "junk" filling up space, so the polymer is softer and more flexible than the ethyl variant which is more compact and harder. The softer variants (which are also safer because they don't decompose to give toxic side-products) are used for sticking humans together – skin cuts (battlefield surgery), gastric bleeding, brain cells. The common general-purpose version is the ethyl because its overall hardness is an advantage.

The second set of differences arise because standard ethyl-cyanoacrylate is low viscosity, ideal for very fine joints and cracks. Other grades add ordinary non-reactive polymers to thicken up the glue to stop it from running too much when larger joints are pushed together, and also to add some toughness. Toughness can also be provided by adding tiny rubber balls, as discussed in more detail in the section about epoxies, where the benefits of the right type of rubber balls produce exceptionally tough structural adhesives.

Some create a gel (for convenience) by adding very fine silica particles.

A third set contain polymerization inhibitors which are great for extending the lifetime of the tube, with the obvious downside of a somewhat slower cure or a need for an activator other than whatever happens to be around in the atmosphere or on the surface.

These activators can be water, dilute alkalis such as sodium hydroxide, or other bases such as amines. A major use of the activators, apart from creating a strong bond even more quickly, is to rapidly solidify any excess that has been squeezed out so it can be easily scraped off for aesthetic reasons. The risks and benefits of using the activator on the surfaces to be glued are harder to balance because too much of a good thing is always a bad thing.

As an example of too much of a good thing, research for this book took me into the world of eyelash cyanoacrylate adhesives which features strong debates about whether "nano misters" which spray fine drops of water are good or bad. They are "good" because the water is an effective activator which attaches the eyelash super-quickly, they are "bad" because they make the adhesive too hard and inflexible.

Scientifically, it would be fascinating to know which variant is specifically good for any given fixing job. If you have to choose a superglue for a regular job then it is possible to carry out the scientific and practical tests to find the best balance of monomer, added polymer, gelling agent, rubber balls and activator. For example, for those who need to adhere to PE, my understanding is that (some of) the amine activators are applied to the surface and left for two minutes before the cyanoacrylate is applied. This suggests that when activator solvents are similar in chemistry to PE (they have a general "petrochemical" odour) they can slightly dissolve the surface and provide a small amount of intermingling between the amine accelerator and the PE to increase adhesion – though this seems to require a specific type of activator such as "toluidine". My expectation is that an aqueous activator will not have such a positive effect.

Alternatively, if we just need a tube to fix whatever happens to break around the house, then a cheap, small pack of ethylcyano-acrylate, which goes solid a month after we last used it, is a wise choice.

What about bungee jumping off a bridge, attached by a few drops of superglue? I recommend that you read the fine print used in one of the many YouTube adverts showing the amazing strength of their glues: *"Professional stunt using backup safety equipment. Do not attempt"*. The key is to stick the bungee cord to the bridge via two super-smooth steel cylinders and ten drops of glue evenly distributed. The forces will be exerted via a pure butt joint and the superglue will be OK.

Similarly, if you really want to lift your pickup truck via a few drops of superglue, attach it via a pair of steel cylinders and drops of adhesive arranged so the truck exerts a pure butt pull. If you try it with any non-vertical loads, involving shear or peel, your truck (or the beams lifting it) will come crashing down.

We cannot leave the discussion of cyanoacrylates without discussing the "fires" when non-inhibited cyanoacrylate is dripped onto cotton. There are a few well-reported cases (I spent a fascinating, though grizzly, hour reading some of these reports) of people being "burned" when they spill, say, 5 ml of superglue onto cotton clothing. There is no disputing that the many –OH groups on cotton are great initiators of the cyanoacrylate poly-merization, and that the polymerization gives off plenty of heat. It is also known that heated cyanoacrylate produces fumes (especially liked in crime scene detective programmes where the fumes reveal fingerprints). The heat is enough to cause burns to the unlucky people wearing the cotton, and the fumes look like fire. But getting cotton to catch fire is difficult. We can be con-fident that most cases of cyanoacrylate "fires" are merely fumes, though the burns are real and very painful.

5.7 EPOXIES AND URETHANES – THE BASICS

Many of us will have bought a 2-pack epoxy adhesive for a challenging adhesive job. Each component on its own is stable indefinitely and they need to be mixed thoroughly before being
applied. Once mixed, the system starts to "cure", i.e. to polymerize.
If the separate components were simple molecules with reactive
groups at both ends, the polymerization would lead to a conventional linear polymer, like our soft spaghetti. While some linear
polymers have great properties (those are the ones we use a lot, such
as PE or PMMA), most have nothing much very good about them as
general purpose polymers or adhesives. That applies also to most
linear epoxies. To be good adhesives they need to have a proportion
of three- or four-functional components that can react away from the
main chain. At first, these side reactions give a "branched" polymer
which would be as useless as the linear one. As the system builds up,
ends of some of those branches react with other branches so that a
complex network of links is created. We now have a crosslinked
system, and that gives significant strength to the epoxy. By altering
the ratio of components, we can create a range of crosslink levels
which, in principle, can give a good range of strengths.

There is an obvious problem with this process. Suppose I want a
really crosslinked system. At some point I have a reactive end of
one branch (in Figure 5.9 this is shown as a semicircle) eager to

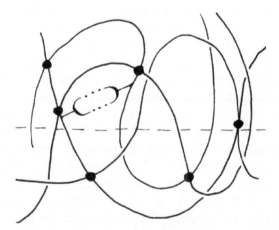

Figure 5.9   After a certain level of crosslinking, it becomes impossible for two
reactive groups (shown straining to get to each other) to react. The
crosslinking process comes to a halt.

react with a nearby part of another chain (the other semicircle). The problem is that the system is already so crosslinked that the reactive end is unable to move the few nm required to complete the reaction. If you monitor the process with some sort of probe that can count the number of unreacted molecules you find that their number decreases rapidly at the start, then reaches a plateau. No matter how long you wait, there is no further reaction.

There is one way out of the problem: heat the system to make it more mobile. When you do this, you see the number of unreacted molecules reach a lower plateau, so you heat some more ... until you've reached whatever balance you want. For many epoxies used in industrial applications (e.g. sticking aircraft wing components) it is necessary to go over 100 °C to get the degree of crosslinking that provides the mechanical strength required for a strong adhesive bond.

A tragic example of an under-cured epoxy is provided by the Boston Big Dig tunnel. We will discuss in Chapter 7 the way to secure heavy loads via bolts into a wall or ceiling. That technique was used to secure heavy ceiling panels to parts of a tunnel. The epoxy was a relatively quick-set variety, convenient but less strong, and without high temperatures the network was incomplete because reactive groups could not move to add another crosslink. Over time the bolts were able to creep within the under-cured epoxy and one panel crashed from the ceiling onto a passing car, killing a passenger.

Because adhesion is a property of the system, strength isn't everything, though for epoxies (certainly for the Big Dig tunnel) it is often one of their key requirements. The adhesives for sticking aircraft wings together have classically been epoxies. Over the years, epoxy designers were able to alter both the amount of crosslinking and the molecular shapes of the reactive components, making ever stronger and stronger epoxies... until they hit a limit. At the limit their epoxies became uselessly brittle. Remember, brittle means that the structure is so rigid (good if you want strength) that there is nothing to dissipate the crack energy, which means poor adhesion.

The solution to this problem was to build in something that could absorb the crack energy: some rubber balls. These at first seemed to be a good idea, then a problem quickly became apparent: because the epoxy wasn't strongly adhered to the rubber,

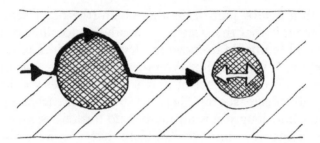

Figure 5.10   A ball of rubber in an epoxy is no barrier to a crack – it just travels around it. But if the rubber is covered with a shell that is integrated into the epoxy the crack stops because the rubber can dissipate the crack energy.

the cracks could often zip straight around the rubber. Even worse, Griffith's law, discussed in Chapter 2, got in the way. The law says that the ability for a crack to open increases steadily as the size of any defect increases. A 1 μm rubber ball, acting as a defect, might allow a crack to form that would not have happened without it.

The solution that has taken epoxies to amazing levels is core–shell rubber particles. The core is the rubber and around it is a shell that can react into the epoxy matrix. Now the rubber can do its absorbing of the crack energy, while the crack does not see any weak links around the rubber (Figure 5.10).

This trick has been borrowed by some cyanoacrylate (super-glue) formulations, as mentioned above and, incidentally, is how PVC window frames are toughened.

Making epoxies work at high temperatures is routine, and their ability to cure better and better at high temperatures has become a potent enabler for many applications. I once, however, came across an ingenious solution to the opposite problem – an epoxy that needed to be handled and cured in an arctic environment down to −40 °C. The formulators developed an epoxy that (with enough time!) could cure at those temperatures, which was a smart first step. The really smart part, though, was fixing a key flaw in the cured material: at those low temperatures it was too brittle, not because it was over-cured (it couldn't be) but because of the cold. Adding those same core–shell rubbers designed for super-high performance, highly cured epoxies that were too brittle at room temperature and above had a dramatic effect on the effective strength of this low-performance epoxy at those low temperatures.

I greatly admire the inventors who could make the imaginative leap between the two very different scenarios, linked by the fact that a brittle structure doesn't know whether it is brittle because it is over-cured or because it is too cold.

One key feature of epoxies highlights a problem with most other adhesives. A standard polymerization process starts with many separate, small, molecules and creates a few large ones. The result is more compact, so the adhesive shrinks during cure. One reason we have to clamp joints during adhesion is to allow the system to adjust to this shrinkage. The epoxy reaction, shown in Chapter 2, opens up the very compact epoxy ring, resulting in an increase in volume that compensates for the normal reduction in volume as the smaller molecules join together. Epoxies, therefore, are used where low shrinkage is a necessity.

Urethanes follow many of the rules of epoxies; they can be two-pack adhesives, though the most popular general-purpose consumer brands are stable one-pot formulations that, like cyanoacrylates, just need a bit of moisture from the air to kick off the reaction. While epoxies are known for their strength, poly-urethanes are often desired for their flexibility and toughness. There is probably no fundamental reason why the two classes have diverged in this way. It seems conceivable to make high-strength urethane and flexible epoxy adhesives. It just happened that the convenient reactive components for epoxies tended to result in high strength adhesives and the typical urethane components tended towards the tough/flexible adhesives.

One of the oddities of polyurethanes is that although they require a *bit* of water to get the reaction going, if they meet *a lot* of water, they start to foam via a reaction that gives off carbon dioxide. The fact that urethane glues can foam seems to me to be a problem. Foams are weaker than solids. Chemicals that, in the rest of the system, are doing a good job at creating the right strength and flexibility are now being converted into gas. I have found literature (which is admittedly biased in supporting their own non-urethane adhesive) pointing out that joints with urethanes can have hidden voids from occasional foam bubbles; these are failure sites for water to enter and cracks to propagate. But one of the rules of marketing is that if there is a fault in a product, convert it to a feature. The "feature" here is that in joints where there is a large gap, the adhesive automatically sees more moisture from the air and therefore foams up – expanding to fill the gap.

The problem of gaps in joints is a severe one. A typical example is re-fixing a wooden joint. As an old joint comes apart, maybe a bit of wood splinters off. Or perhaps the wood has shrunk over time creating a large gap around the piece you have to re-glue. So far, my descriptions of adhesion have assumed near-perfect contact, with the properties of the adhesive itself not so important on their own, and very important in terms of the system. As soon as we have gaps having to be filled by the adhesive, the general mechanical properties of the adhesive become more important. Although the literature does not have much to say on the mechanics of gap-filling foams within products that are being sold as adhesives, it seems to me that foam is unlikely to be superior to a solid epoxy – as an adhesive – though it is debatable whether it is better or worse than a cyanoacrylate.

There are urethane products that are designed to produce lots of foam – these are the foam fillers. The foam fills large spaces (Figure 5.11), sets hard and provides some modest adhesion across the filled gap. Such foams can be the preferred option. There are no doubt plenty of cases where it is desirable to have a product that is designed mainly as an adhesive and which can foam into gaps. We just have to be careful not to take that

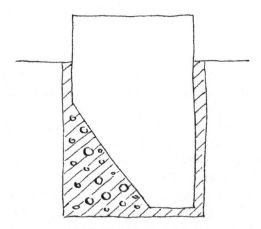

Figure 5.11   If a urethane adhesive creates a foam in a big gap this might be good, because it fills the gap, or bad because it is filled with a weaker material.

foaming "feature" as to be something that is desirable in all the applications of this type of adhesive.

There is one aspect of urethanes that is both well-known and ignored in practice. Urethanes, by their nature, contain "aromatic isocyanates", known as MDI and TDI. These are well-known to be toxic, allergenic and suspect carcinogens. If they were invented today, they would not be allowed. If some evil chemical company tried to introduce them, there would be protests outside their headquarters. Why, then, are they used in just about everything: home adhesives, automobile adhesives, shoes, food packaging and more? Because polyurethanes are wonderful materials for which there is no obvious substitute. In food packaging applications, the trick is to say that there is no MDI or TDI in the final product because the amount of the other reactant (a diol) is sufficient to guarantee 100% consumption.

This "there is no alternative" approach to safety has to be matched with "there is no evidence in practice that these things cause harm". If the second part was more generally adopted, then we would have far fewer scares about safety. A wonderful example of a nasty compound used in large quantities on a daily basis is paraphenylenediamine used in hair dyes. This is another compound which, if introduced today for large-scale consumer use would instantly attract zillion-dollar lawsuits. The problem is that there is no known satisfactory alternative; if it were withdrawn, the hair dye industry would collapse. To avoid this, and because hair dyes are very popular, the regulatory agencies permit their sale with suitable restrictions and warnings, and large quantities of this material are used.

So, in this spirit, carry on using urethane adhesives whilst following the manufacturer's safety instructions. My pot of urethane glue has plenty of warnings about things such as skin sensitization or asthma, so users have no excuse for not taking adequate care. They are great adhesives and the risks to you from the MDI/TDI are minimal and manageable.

5.8 EPOXIES AND URETHANES – LOCKING ON

If you apply an epoxy or a urethane directly to most plastics and to many metals, you have a wonderful adhesive product... with very little adhesion in the system. There is no chance for intermingling

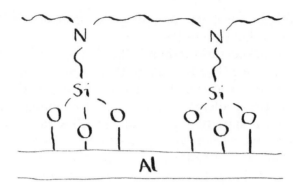

Figure 5.12   Repeating 5.7, an adhesion promoter such as APTES can grab on
to the aluminium via the siloxane bonds while the $-NH_2$ at the
other end can integrate into an epoxy or urethane network.

and entanglement via polymer compatibility, and there are no
suitable reactive groups at the surface of the adherend to take part
in the epoxy or urethane reactions. The adhesion is poor because
there is no entanglement across the interface.

Turning that problem around, if reactive groups can be made
available then strong adhesion becomes straightforward. In
some cases, they are naturally available in the forms of hydroxyl
(–OH) groups which are able to react, if somewhat slowly, with
epoxy or urethane systems. Far better are amine groups, $-NH_2$
which react very well into the adhesive system (Figure 5.12).

This is where adhesion promoters, or primers, come into play.
For aluminium, for example, the trick is to use APTES, referred
to earlier, which has the silane "-TES" part that sticks well to the
metal leaving the amino "A-" part ready to react with the epoxy
or urethane. [The propylene -P- is the squiggle in the diagram
that joins the two ends together.]

The more general power of the silane "-TES" part will become
apparent later.

5.9   UREA/MELAMINE FORMALDEHYDE RESINS

A different type of reactive adhesive is based on the way that formaldehyde reacts with molecules containing multiple amine groups
such as urea (two amines) and melamine (three amines). These
are the urea-formaldehyde (UF) and melamine-formaldehyde (MF)
adhesives. With careful control, a manufacturer can make a dry

powder that contains a partially reacted system with, importantly, no (or, rather, insignificant amounts of) free formaldehyde, which is a gas at room temperature. When this powder is mixed with water, the polymerization reaction can continue, creating a very hard cured product. At room temperature the reaction is relatively slow, which is good for those who need plenty of time to reposition things and have the right clamps to keep things together until the reaction is complete. Those who want a faster cure can raise the temperature and/or add an accelerator/catalyst, typically an acid.

The two-part Aerolite adhesives from the 1930s were revolutionary in terms of creating water-resistant adhesion for aircraft and boats; typically the first part, the urea/formaldehyde powder was mixed into a syrup with water and pasted onto one surface as a typical slow-curing UF resin. The second part, the acid catalyst, was pasted onto the mating surface. When the two surfaces were brought together, the cure was rapid, allowing fast construction of complex shapes, such as wooden aircraft. It might seem odd today that pilots were happy to fly high-performance Mosquito aircraft, knowing that they were just stuck together with Aerolite 306, but it is a reminder that the right adhesive applied in the right structures can manage the balance between strength and brittleness. The system is still approved for wooden aircraft construction.

For wooden joints, the system can react around and with the fibres, creating a strong interfacial bond. The adhesive itself is strong so the overall system is strong. The reader should, by now, be alert to the potential downsides. These bonds are brilliant in, say, laminated structures where strength/hardness is paramount. They are less good in typical household joints such as chairs that are subjected to lots of flexing, because "strong/hard" has the downside of "brittle".

The big marketing downside of these adhesives is, of course, that the word "formaldehyde" is popularly associated with health risks. All reactive adhesives by definition contain reactive chemicals, so they all pose some sort of risk. Those who choose not to wear gloves and not to have adequate ventilation create their own risks. With common sense the risks are minimal and unless you fill an ill-ventilated room with lots of UF/MF adhesive joints, the long-term levels of formaldehyde are not a problem. It has long been known that formaldehyde is commonly found in natural materials; you can get plenty of formaldehyde emitted from, say, pine panelling in a home. But formaldehyde is now so demonized, rational arguments about relative risks are not possible.

Rather than focus on the chemical risks, it is better to focus on the risks of a chair collapsing unexpectedly from a bad joint or (because UF/MF is used extensively in a marine environment) your boat sinking because the wrong adhesive was used.

5.10   WILL NOT STICK TO PE/PP – OR PET

PE and PP are cheap, strong, tough polymers which are inert to most solvents and chemicals, impervious to water and readily formed into bags, blocks and, in the case of PP, large chunks of cars such as bumpers/fenders. They are "polyolefins" which means that they contain only carbon–carbon bonds.

Their inertness is a great strength except when it comes to sticking to them. They have no reactive groups at the surface. They are also highly crystalline which means that normal solvents cannot get into their surface to allow polymer-to-polymer intermingling or entanglement. This is inconvenient in the home because if you carefully read the labels on many Super Strong Adhesives, they admit that they will not stick to PE/PP. It is even more inconvenient for industry, where sticking to PE and PP is of huge importance, as an example from the packaging industry illustrates.

PE and PP are wonderful polymers for food packaging. They don't let water in or out of the package and block escape routes for many flavour chemicals. However, the "simple" PE packaging we are so familiar with is not at all simple (Figure 5.13). PE is a very poor barrier for some flavour molecules (such as limonene, a strong citrus flavour) and a poor barrier for oxygen. Because of these barrier limitations the PE packaging film contains a middle layer of a polymer called EVOH which is a poor barrier for those things (such as water) where PE is a good barrier, and a great barrier for those things (such as limonene or oxygen) where PE is poor. In making these "simple" bags at least five layers of polymers have to be squirted out ("extruded") together: PE, PE-tie layer, EVOH, PE-tie layer, PE. The "tie layers" are intermediate polymers that can bond with the PE and with the EVOH which are mutually incompatible.

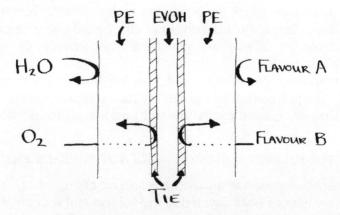

Figure 5.13 A "simple" packaging film has at least five layers. The outer PE is a good barrier to water and some flavours and bad for oxygen and other flavours. The inner EVOH is the opposite, bad for those things that PE is good for and good for the bad things. Between them they stop most things from going through. The tie layers keep the two polymers together.

A further downside of the inert nature of PE and PP is that the apparently simple task of printing onto food packaging is basically impossible onto standard PE/PP. And when it comes to creating sophisticated packaging such a boil-in-the-bag packs held together by urethane adhesives, again there is no way to get PE/PP to stick. Before finding out how we get things to stick to PE/PP we need to see a similar problem with a different polymer, PET.

Many of us drink milk from PE bottles. The milk lasts for a long time and we don't care that the bottle itself is not transparent. When it comes to more fancy drinks, PET is the polymer of choice. First, it is see-through, which is good for marketing purposes. More importantly, it is, unlike PE, a good barrier for CO_2 so fizzy drinks stay fizzy, and it is a good barrier to many flavour molecules, so the drinks stay tasty. For packaging applications that require higher temperatures, PET is also preferred as it has a much higher melting point than PE. The problem for all those using PET is that it is just as impossible to stick to as PE. Unlike PE, it has plenty of chemical functionality (because it is an ester of ethylene glycol and terephthalic acid) which should in theory make it not too hard to stick to. The secret to its resistance to sticking, its non-printability as well as to its success as a packaging polymer is that, like PE, it is highly crystalline. Even an ink/solvent that should be chemically compatible is unable to penetrate the surface to create entanglement.

To reach our destination of describing how to get things to stick to the polyolefins and to PET we have to make a detour into topic that has caused confusion and puzzlement for decades.

5.11 POLYOLEFIN ADHESION – IT'S NOT SURFACE ENERGY

There are many ways to measure the surface energy of a polymer. There are whole books devoted to the subject and they are, sadly, rather a waste of time. First, because many of these methods turn out to be inaccurate. Second because, as we found in the second chapter, surface energy isn't all that important for adhesion. In the context of polyolefins, that second reason is almost a declaration of war, on two fronts. I have enjoyed the battles over the years because no matter how hard anyone tries, they cannot fight the laws of physics, and in both these battles, the laws of physics are on my side.

If you are unaware of the passions around these battles, then I hope you will find that what I am writing is rather straightforward and obvious – which it is. If you are on the opposing side, well, apologies in advance, this straightforward and obvious explanation is the correct one.

Let us start with the smaller of the two battlefronts. I say that surface energy is not important for adhesion and immediately someone will say "Ah, but you cannot get good adhesion if the adhesive doesn't wet the adherends". Although that statement is true it is either trivially addressed or is provably irrelevant. It *is* tricky to coat an adhesive onto a polyolefin if the coating solution does not adequately wet it. But just about any coating formulator can solve this problem through the right choice of solvent or additive (such as a volatile surfactant for water). More often the criticism is meant for those systems where the adhesive is laminated between two layers in a sequenced process; yet here the criticism is irrelevant because *all* adhesive types will wet out on *all* possible combinations of adherends. The theory behind this fact has been known for a century, yet it is often unknown to those who write books or chapters about surface energy. We don't need to get into the theory because we can see what happens with a very simple experiment.

https://youtu.be/w1dsiMOvdRc

Take a polyethylene bag – say a sandwich bag – and cut it so you have two single sheets. Put a drop of water onto one sheet, maybe, as I've done in the video, colouring the water with food dye to make it easier to see. The water does not wet polyethylene; it sits there without spreading out. This is the first sort of non-wetting. Now bring the second sheet near the drop, holding it with your fingers away from the centre as a loose sheet without tension. As you bring it close to the water, the sheet jumps into contact and the water rapidly spreads out. Yes, it spontaneously wets the two surfaces in exact contradiction to the claims of the surface energy people.

The explanation is simple. If the water has a choice between spreading on the first sheet or sitting as a droplet, it chooses to sit. The water has no great affinity to either air or polyethylene, so it reaches an uneasy compromise. Now bring the second sheet close to it. The water has to choose between air or polyethylene and now it is no contest; although the polyethylene is not much liked, it is far preferable to air. In scientific terms, the surface energy of polyethylene is low, and therefore not much liked by water, but it is much higher than that of air.

Now to the major battlefront. If you pass your polyolefin through the right type of flame or you hit it with a high voltage system called a corona or with a somewhat different system called a plasma, two things happen: the surface energy increases *and* the adhesion to many materials increases too. Using a form of logic identified by Aristotle as being faulty and given the Latin phrase *"post hoc, ergo propter hoc"* ("after this, therefore because of this"), the surface energy proponents say that adhesion increases because of the increase in surface energy.

For these polyolefins, a typical treatment will raise the surface energy from 32 to 42 $mJ\,m^{-2}$, an increase of 30%. The measured peel, however, can go from, say, 32 $mN\,m^{-1}$ to 100 $N\,m^{-1}$

(i.e. $100\,000$ mN m^{-1}!), an increase that is completely off the scale compared to any surface energy effect.

More extremely, PET starts with a surface energy of 42 mJ m^{-2} and nothing will stick to it. After treatment it might just about rise to 50 mJ m^{-2} – and adhesion will be rock-solid. Such a leap in performance is inexplicable via surface energy. Another example breaks the *post hoc* "surface energy-therefore-adhesion" argument chain completely: if you zap the PET with the powerful light of a xenon flash, the surface energy stays unchanged, yet adhesion can leap to rock-solid.

When we disentangle all this, we find that there are two effects (Figure 5.14).

1. Entangled adhesion becomes possible. The problem with the polyolefins and with PET is that they are highly crystalline. If you try to apply solvents to encourage other polymers to cross the interface and get entangled, nothing happens – the crystals remain crystalline. The blast with the flame, corona or plasma wrecks the surface, reducing the crystallinity, making it easy for the right polymers to intermingle using the right solvents. Because too much of a good thing is a bad thing, over-treating the surface reduces it to rubble (with a high surface energy, which looks deceptively good for adhesion) which indeed sticks to the other surface but has little cohesion to itself, and any bond onto it will fail. For pure PP, there is no safe level of corona treatment – it gets

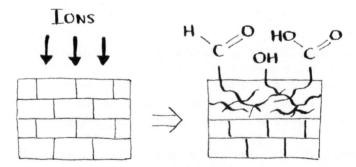

Figure 5.14 A crystalline polymer surface when bombarded by ions from a corona or plasma becomes open for entanglement and, at the same time, acquires some chemical functionality which may or may not be useful.

reduced to rubble when you have the minimum energy needed to excite it. For PET the xenon flash is wonderful. It heats the surface super-fast to a super-high temperature, which melts the polymer, which then cools super quickly into an amorphous form that is easy to stick to.

2. The treatment (though not the xenon flash) adds some (a few %) chemical functionality, typically hydroxyls, carbonyls and carboxyls. If your adhesive reacts nicely with any of those functionalities, then this few % is exactly what is needed for strong entangled adhesion. As it happens, these functionalities are seldom good enough for epoxies and urethanes which react with themselves faster than they react with the functionality on the surface. To get strong adhesion via this functionality we need another trick discussed next.

Too little treatment has too little effect. Too much treatment creates rubble at the surface which will give unreliable adhesion. Although millions of m^2 of good product are produced every day thanks to these surface treatments, the problem is that with the wrong explanation for the beneficial effects of the treatment, it is hard to find the optimum treatment and to diagnose problems when they occur.

There, the battles are over. Surface energy has been defeated and we can get on with the rest of the book. We can celebrate with a *very* interesting story of how we can get great adhesion with unpromising components.

5.12 ADHESION VIA THE WORLD'S WORST POLYMER

Polyethylene imine, PEI, is a polymer with almost nothing good to say about it. It has a bunch of "secondary amines" along the chain and these give it water solubility. By making an even worse polymer, with side branches, we end up with a PEI containing lots of primary amine ($-NH_2$) groups (Figure 5.15), the ones that love to react quickly with epoxies and urethanes and also, as it happens, with carbonyls and carboxyls. We have the unlikely elements for a truly excellent adhesive system.

It starts with a polyolefin (PE or PP) suitably treated to have a number of hydroxyl, carbonyl and carboxyl groups. The hydroxyls are useless but mostly harmless. We will forget about

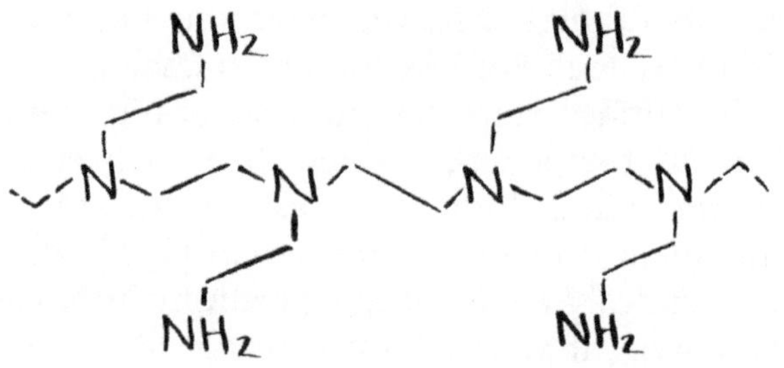

Figure 5.15   This is a PEI, polyethylene imine, with plenty of extra $-NH_2$ groups ready to be reacted as part of a complex adhesion promotion process.

them. Now coat the polyolefin with the PEI. Some of its primary amines react quickly with the carbonyls and carboxyls (forming imines and amides). We now have PEI stuck to the polyolefin. The PEI still has plenty of unreacted primary amines. Now we can coat the surface with our epoxy or urethane, the primary amines react into the adhesives as they cure and we almost have rock solid adhesion (Figure 5.16).

The remaining problem is that PEI is a useless, weak polymer that can also attract water to the interface, able to do mischief over time (for example, destroy the imine and amide bonds). If you have 100 nm of PEI between the polyolefin and the adhesive,

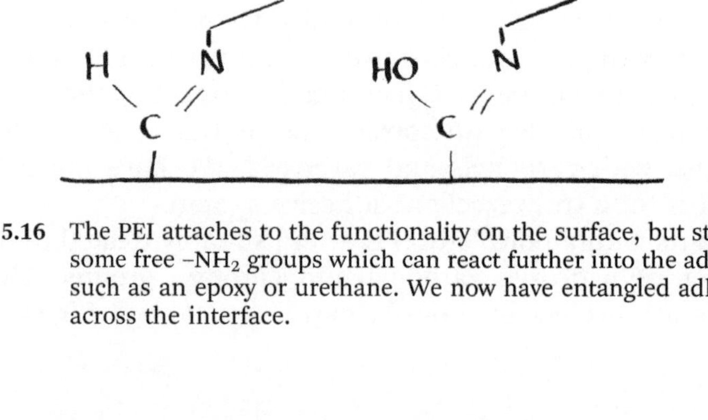

Figure 5.16   The PEI attaches to the functionality on the surface, but still has some free $-NH_2$ groups which can react further into the adhesive such as an epoxy or urethane. We now have entangled adhesion across the interface.

any attempt at a peel test will fail because the system breaks along the weakest link, the layer of PEI.

If, however, you can coat 20–50 nm of PEI, you get all its benefits, with a coating so thin that the crack energy does not rip it apart and where enough of the amines have reacted to reduce the amount of water being attracted to the interface.

Millions of square metres of olefin/urethanes are produced each year using the PEI trick – plenty of the packaging in your house contains PEI. Providing the right PEI (suitably branched and also lightly crosslinked to give it an extra bit of strength) is a good business for the few companies who have learned how to produce the right product. It has always amazed me that anyone could invent such a system; whoever did so has my full admiration.

5.13 UV ADHESIVES

If at least one surface is transparent, an alternative form of adhesive works very well. The adhesive is a somewhat sticky and somewhat smelly combination of 2-, 3-, 4- etc. functional acrylates, i.e. vinyl (carbon–carbon double bonds) groups with carboxylic acid group that can be attached to a simple or complex array of alcohols to create these multi-functional esters. Like all vinyl systems they can be polymerized using radicals from the breakup of molecules such as benzoyl peroxide. Some forms of fillers/caulks are based around such a system as is one type of nail polish.

Here we are more interested in the fact that formulators can add a small amount of a special type of molecule, a photoinitiator, which absorbs UV (or "blue") light. The energy from the absorbed light causes the photoinitiator to break down into radicals, kicking off the polymerization reaction.

Provided that the UV light can reach all the way through the system, curing can take place in a fraction of a second. If one surface is exposed to oxygen, which is a potent inhibitor of the polymerization reaction, it is harder to get a good cure. Consumer-grade UV adhesive "pens" have been around for a while. They come with a blue LED curing light. As theory suggests, and YouTube videos show, for transparent joints they can be quite effective, for joints where the light cannot penetrate very far, they are of little use.

Because there is a limitless array of acrylates, the formulator has a large choice of what to include in the adhesive system. Lots of small molecules give a (relatively) low viscosity adhesive that can be solidified rather quickly into a hard-crosslinked system. Lots of polymers with acrylates attached are (much) more viscous to start with and give a more flexible result because the polymer makes up the bulk of the adhesive. All possibilities in between can be arranged. Indeed, the problem for the formulator is that the choice is too wide and there is always that hope that one more formulation tweak will give even better results.

Getting a good crosslinked adhesive is straightforward. How do these systems also gain entanglement across the interface for strong overall adhesion? In some cases, the liquid acrylate system can be chosen to intermingle with the adherend surface then get entangled via the crosslinks on cure. If the surface does not allow such soft interactions, then it needs to supply the appropriate functionality. The amines of a PEI surface react rapidly with acrylates just as they do with epoxies and urethanes, expanding the range of the versatile PEI system. For adhesion to aluminium or glass, an acrylate-TES system works well and for steel an acrylate-phosphate system is chosen. With ingenuity, most formulators can find a good way to get UV acrylate adhesives to stick well to most surfaces.

With UV systems, we encounter a problem related to the one we met with epoxies, urethanes etc: as the system cures, it gets more difficult for reactive groups to move the short distance to react with an awaiting bond. This means that the cure comes to a halt rather quickly. The chemically curing systems can carry on curing over hours, days, even months, or can be fixed with a short exposure to a hot curing oven. The UV systems basically stop once the UV light has gone (there are some remaining radicals but these tend to get annihilated over time). This all means that many UV adhesives are not as strong as they might be. The solution, whenever possible, is to cure at a higher temperature so that the reactive groups stay mobile for longer.

There is one final trick. It is possible to make a typical adhesive tape with its soft, weak adhesive that happens to contain some acrylate groups (and a photoinitiator) which have no impact on the tape's adhesion. It acts like any normal adhesive tape and will, for example, stick nicely to a silicon chip to allow it to

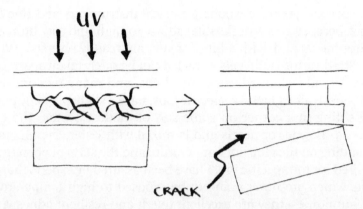

Figure 5.17 A relatively soft PSA sticks nicely to a silicon chip surface to protect it during a production process. A blast of UV makes the PSA brittle, and the film peels off easily, allowing the chip to be further processed.

be transferred to some other process. Once that process has been completed, the chip needs to be unstuck from the tape. This would normally be difficult. With these special tapes, a blast of UV light converts them from soft and sticky to hard and brittle. It then takes little effort to detach the chip from the cured tape (Figure 5.17). This trick is performed on millions of chips in vast semiconductor plants – and the cured tape is discarded as trash.

5.14 HOT MELT ADHESIVES

We all want our glues to have infinite open time, while we try to get the parts together properly, and instant cure – when we've finally got them together. This is impossible. One way to get close to the ideal is by using hot melt adhesives. They are squirted from a gun or, if they are some of the old-fashioned collagen (hide/ hoof/fish) glues, taken on a brush from a hot glue pot, and stay liquid long enough for us to sort out the joint, before setting solid.

Some of these adhesives are designed to be cheap, simple and poor quality. A melt of pure PE when dabbed onto one flap of a carton provides more than enough adhesion when the other flap is pressed into contact. When we want to open the carton, the brittle bond is easily broken.

Others provide strong, but not too strong, adhesion that is readily reversed. These are the hide glues that, as discussed earlier,

are especially good for wooden musical instruments and fine furniture because they are flexible, adjust to high and low humidity
environments and, with a bit of steam, are readily reversed. When
your Stradivarius violin gets a crack it can be safely taken apart to be
fixed. All old-fashioned carpentry shops, and many modern ones
that choose hide glues for specific jobs, have a slightly smelly pot of
glue sitting in a corner on some sort of modest heater. The glue
comes as powder or pearls and is mixed with water, the mix ratio
depending on the glue pot temperature and the sorts of open times
required for the specific job. I have been assured by experts that for
joints which are never going to be exposed to high temperatures
and humidities, they are excellent tough and resilient adhesives.

Another class of hot melts behaves like the pressure sensitive
adhesives discussed in the next chapter. The trick is to use
weaker polymers which, at high temperature, show the normal
"tack" of a pressure sensitive adhesive, while being somewhat
(but not too much) more solid at room temperature.

It is also possible to combine a meltable system with one that
can react and cure once it has been squirted from the nozzle.
This in principle gives a fast bond (by cooling and solidifying) for
an effective quick fix, followed by a stronger bond once the
system has fully cured. For example, the hot melt urethanes
solidify after a reasonable open time (e.g. 60 seconds) and then
cure, like their room temperature equivalents, via the moisture
they picked up while the joint was being assembled. For those
who know what they are doing and who have worked out that the
balance of quick fix and long-term strength is required for a
given type of job, this sort of system is obviously attractive.

5.15 WOOD GLUES

Many years ago, the first wood glue I used was a white emulsion
of PVA – polyvinylacetate. PVA is insoluble in water, so it is
provided as micron-scale spheres of PVA polymer surrounded by
surfactants that keep the spheres apart. As the water dries, the
spheres start to flow together ("coalesce"), forming a solid film
that is water resistant. That was then. Now a typical wood glue
is – a PVA emulsion. In almost every other part of adhesion
science, we have glues that have become more advanced. PVA
emulsions, though, have stood the test of time.

PVA itself is not an especially good polymer and merely wraps itself around the wood fibres to create a reasonable bond. I could readily imagine a number of polymer systems that would be far more effective. They would form a tough polymer *and*, unlike PVA, they would be able to create the right level of chemical reaction with the chemicals in wood. When I looked into the academic literature, I found that my analysis was correct and there are indeed plenty of wood glues that give a far stronger bond when measured by standard tests. Why, then, do we still have PVA wood glues?

The answer, as we find so often, is that too much of a good thing is a bad thing. If we could keep our wood in a controlled humidity environment, the alternative glues would be superior. In real life, humidity changes and the wood expands and contracts. The superior adhesives can be too inflexible to accommodate such changes, and "strong" joints tend to crack and fail over time. The PVA is sufficiently flexible to accommodate the changes, stresses don't build up, and the joint survives.

Why do PVA glues have a distinctive smell? The coalescence of the emulsion particles is necessary to create a good cohesive film. The particles have been protected to stop them from clumping together. The downside of this protection is that there is a resistance to film formation. The glues therefore contain a "coalescing agent", a slowly-evaporating solvent that is a small (insignificant) percentage of the initial formulation and becomes a larger percentage as the water evaporates. This solvent allows the PVA particles to flow together to form a film before it, too, evaporates. A typical coalescence agent, mentioned in patents, is butyldiglycolacetate and it is probably something like that which creates the distinctive odour.

5.16 GENERAL PURPOSE GLUES

In Germany, the generic phrase for sticking something together is to "UHU it". This is a classic case of a trademark becoming a general word, such as, in English, to "hoover" is to use a generic vacuum cleaner. UHU is one of a number of generic solvent-based glues which work adequately if the liquid adhesive can flow into pores or around fibres and if the residual solvent has a chance to diffuse away. UHU (you can find the ingredients online) is based on

PVA and cellulose nitrate and uses a mix of esters (e.g. ethyl acetate), ketones (e.g. acetone) and alcohols (e.g. IPA) to get the right balance of evaporation rate for ordinary users – neither too fast nor too slow. The PVA is dissolved in the solvent so there is no need to compromise its performance in the way the emulsion stabilizers compromise the aqueous formulations. There is no opportunity for chemical interactions across the interface so these glues just provide the perfect surface contact that gives adequate adhesion if the joint isn't subjected to peel forces.

Having written that paragraph I had to challenge my own logic. Playing with some of these glues I realize that they don't dry to form a hard solid. Instead they are relatively soft so they are acting in some ways like a PSA – they seem to absorb crack energy by dissipation. In peel, therefore, they are more robust than I had expected. Because they gain their surface contact by being solutions rather than as solid PSAs, they can be tougher polymers than a typical PSA so the adhesion is somewhat better. This comment will make sense when we find out about the Dahlquist criterion in the next chapter.

As is universally noted, any change away from these solvent-based formulations to more child-friendly formulations results in a significant deterioration of performance. Child-friendly adhesives with water-soluble polymers (e.g. starches or PVP, polyvinylpyrollidone) work in the same way as the solvent-based versions. However, it is hard to find a water-soluble polymer with the adhesive properties of, say, solvent-based PVA. And unlike solvent-based glues where the solvent can fully disappear, in the real world there is plenty of ambient water to stop the adhesive from fully drying out, so the adhesion is readily undermined. For a children's adhesive this might be a good thing, as easy reversibility is sometimes necessary. When adhesive suppliers try to formulate "chemical free" adhesives for adults, the compromises inherent to water-soluble adhesives, especially the ambient water vapour, seem insurmountable.

5.17 SILICONES

I have been in production environments where even saying the word "silicone" is enough to cause people to glare angrily at you. If you were seen with any form of silicone you would be picked up and thrown out before you brought production to a standstill.

Silicones have the reputation of being the ultimate anti-adhesives, and the thought of molecules of silicones floating around in a production line that relies on good adhesion is the stuff of night-mares for production staff. Silicones rapidly come to the surface of any coating or adhesive and produce a low surface energy which can cause problems if you want to apply, say, another coating or paint; they also get in the way of polymers crossing the interface to entangle. We will see in the next chapter that silicones really can produce wonderful, and necessary, low adhesion for, say, release coatings, though not because they have low surface energy.

Yet the International Space Station is stuck together with plenty of silicone adhesives. They provide strength, toughness, flexibility and an astonishing ability to survive the cold of night-time space and the full heat of the sun.

When you have to add silicone sealant to your bathroom or kitchen you know that it will stick fairly well to most surfaces and will also be rather hydrophobic, repelling water and thereby discouraging moulds that like to gather in damp corners – though most sealants now include anti-fungal agents because untreated silicone sealants will eventually go black with mould.

You apply the sealant by squeezing it from a tube and, if your technique is much better than mine, you get a smooth glossy surface which over time becomes solid, sometimes giving off the smell of vinegar while it solidifies. The tube contains a mixture of medium-weight silicones, some of which end with a free hydroxyl group and some of which end with a group containing the basis of acetic acid, vinegar. Amazingly, they don't react in the tube. Once out of the tube, the water vapour in the air dif-fuses into the silicones and, with the help of a tin catalyst, this triggers the cure. The hydroxyls kick out the acetates, creating the vinegar smell (others use "oximes" rather than acetic acid groups, giving off a musty smell), allowing the smaller chains to join together into a solid polymer. Some of the silicones contain multiple groups to provide sufficient crosslinking.

For those of us who cannot create a perfect bead of silicone adhesive the first time, and who find that an attempt to run a finger along the bead causes more problems than it solves, I have been advised on two techniques. The first is to use masking tape to define the edges. The light, pressure sensitive adhesion of such tapes is described later. The second is to put a drop of dishwashing liquid onto your finger. The liquid doesn't stick to your finger and the silicone doesn't stick to the liquid.

Although these silicones are often intended as sealants, they can provide impressive adhesion. If an old-fashioned ceramic bath has been "set" into a bed of silicone sealant, ripping out the bath years later can be a challenge. Part of their strength comes from the flexibility of silicones. Cracks are readily dissipated by the rubbery silicone. The other part comes from the complicated chemistry. In Chapter 1 we saw the rather confusing terminology around "sil-" chemistry, with silica, silicates, siloxanes etc. It is the "silane" part of the formulation which provides bonding within the adhesive and the entanglement at the interface. We shall come to that in a moment.

Because silicone chemistry is so versatile, it is possible to create genuine silicone adhesives. At one extreme, as mentioned, the International Space Station is stuck together with silicones. At another, Type A Silicone (a fancy name for a rather conventional silicone) is certified for use as an adhesive/sealant for medical devices such as pacemakers.

https://youtu.be/hTwnZo0IJo0

There is a safe and rather impressive silicone adhesive that we can all get hold of (literally, it is safe to use with your hands). It is called Sugru and is a fairly thick paste of silicone containing (brightly) coloured fillers. It is wrapped in an aluminium-lined pouch to keep water away from it during storage. When you open it and mould it with your fingers, that helps to provide the moisture needed to kick off the polymerization. After a day or two (depending on thickness, levels of moisture etc.) it turns into a semi-flexible solid with rather good adhesion to many surfaces. As you can see at the end of the tape test video it sticks to glass and ceramics, metals, wood, many plastics (but, as discussed later, PE and PP are a challenge) and more. Reading the Sugru patents (I spend a lot of time reading patents – there is a lot of good science to be learned from them, though you cannot believe everything you read in them) we can see examples of how they can get good adhesion not just through the standard silicone tricks but also via adhesion promoters.

The patent suggests the use of GPTES where the G stands for "glycidyl" which is another word for epoxy. You see how adhesion science comes together so nicely! We know that epoxies can react slowly with hydroxyl groups on surfaces, and most surfaces have some hydroxyl groups, so epoxies can give adequate adhesion on many surfaces. The G of GPTES reacts with the low levels of hydroxyls on the surface and the -TES part (joined to the G via the Propyl group), integrates into the overall silicone and we end up with good adhesion.

Silicone sealants and adhesives contain plenty of the "-TES" that have been referred to a number of times, along with "-TMS". As these are the key to many strong adhesives (not just silicones) it is time to discuss them properly.

5.18 TES AND TMS – WHY SO MANY ADHESIVES CONTAIN SILANES

TES is "triethoxysilane" and TMS (Figure 5.18) is "trimethoxysilane". Each silicon atom has four bonds. One of the four is either part of the adhesive or is a group, for example a propylene spacer group (P) ending with a suitable functionality X that can react further, such as the amine in APTES with urethanes and the glycidyl group in GTPES which reacts with –OH groups. The remaining three groups can be the alcohols ethanol or methanol,

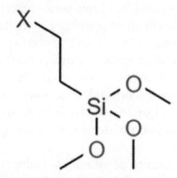

Figure 5.18   This TMS, trimethoxysilane has three groups that can react with –OH groups on a surface or within a polymer, and a functionality, X, which can react further into the system.

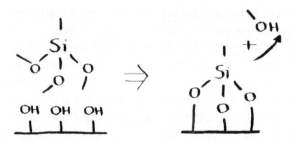

Figure 5.19 The three methanol groups attached to the silane can readily react with –OH groups on a metal oxide surface such as alumina, giving off three molecules of methanol, and firmly attaching the silane to the surface.

or they can be –OH groups which technically make them "silicic acids" though, fortunately, these don't hang around for long.

The great thing about these three-group molecules is that just about any surface containing just about any –OH functionality can displace one or more of the groups, creating a strong bond to the surface (Figure 5.19). When one adherend is aluminium we are in fact trying to adhere to some form of aluminium oxides/ hydroxides at the surface, and the result is a hybrid silicate/ aluminate. With those oxides and hydroxides in play, adhesion to aluminium is strong – a common trick for the aluminium used in multilayer food packaging. Most of the ceramic materials used in baths (the old-fashioned ones), bricks, blocks, plaster and cement are silicate/aluminate, with surfaces that are well-tuned to react with TES/TMS. Wood has plenty of –OH groups ready to react. Many polymers have some reactive functionality at their surfaces either via their specific chemistry or from natural oxidation, so they too can stick to the silanes.

The only other type of general-purpose adhesion group available is the carboxylic acids in some acrylate adhesives and in those where special dicarboxylic acids such as maleic acid have been incorporated (Figure 5.20). Their uses are more specialized because their bonds usually do not have the general robustness of the silanes.

Returning to the theme of silicones, there is a recent trend in adhesives that make use of the TES/TMS part of the silicones and do away with the silicone part. One of the strengths of silicones, their inertness, makes them a nuisance because they are very

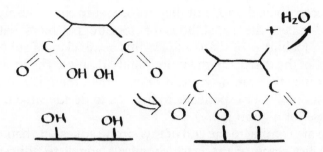

Figure 5.20 Maleic acid can be incorporated into polymers, ready to attach to any surface with –OH groups. The fact that it contains two carboxylic acids in close proximity allows it to "chelate", grab on with two pincers like a crab's claw. This makes it generally more robust than using just one carboxylic acid.

difficult to paint. Also, as anyone who has handled them knows, they can readily spread and cause problems outside the bonded area. Their flexibility, which is such an important plus for silicones can be a limitation when a more solid joint is required.

The newer class of adhesives has a variety of names such as MS or SMP, with S representing silicone and M being something like "modified" and P being polymer. They are also known technically as silane hybrids. If a product doesn't have an obvious S-based name, the clue that it belongs to this class is that they are happy to be associated with (or can be applied under) water (which is necessary for their cure) and they don't claim to be any of the obvious types such as silicone, acrylic or urethane. The ingredients lists often say that they contain some of the "sebacate" and "vinylsilane" additives mentioned below.

These silane hybrid adhesives take relatively short chains of polymers with desirable properties such as a polyphenylene oxide, an acrylate or a urethane, and put a TMS group on each end. In the presence of water and with a suitable catalyst to speed up the reaction, bonds are formed across TMS groups, creating a high molecular weight and, with suitable additives, crosslinked polymer. The TMS groups also adhere well to the oxygen functionalities at the surfaces of the bricks, metals, woods etc. The adhesives emit small amounts of methanol during curing and some of the safety data sheets mention this because methanol is unpleasant at high concentrations. TES-based versions emit the more acceptable ethanol, but they are slower to cure. The ability to mix-and-match the capabilities of specific polymers with the general capabilities of silanes allows this newer class of adhesive to fix some of the deficits in the more standard range. The problem is that because they are newer and more varied, there is not yet the background of user experience to help us to decide which, if any, to choose for a given job.

As we are coming to the end of this strong adhesion chapter, it is time to list some of the extra chemicals added to adhesives to provide an optimal set of properties. Some people absurdly want "chemical free" products which is impossible because all products are made from chemicals; water is as much a chemical as any other. Others claim that evil corporations add toxic chemicals for no good reason. This too is a nonsense. A really good way of going out of business is to kill your customers. Industry spends far, far more time worrying about chemical safety issues than consumers do and are forever having to re-formulate products because some necessary chemical (such as a biocide to stop sealants from growing mould) which was previously deemed safe has now been deemed unsafe, often on the basis of the flimsiest of evidence. The "precautionary principle" works in both directions. Of course, if there looks to be a reasonable doubt about a chemical, it is a good precaution not to use it. But we should also take the precaution of not throwing away too many chemicals that provide key functionalities for things like adhesives.

5.19 "CONTAINS COMPLICATED STUFF"

Some readers, will, like me, be fascinated by the ingredients listed on adhesive packs or in safety data sheets. Those who aren't so interested can feel free to skip to this chapter's short final section on the ideal strong adhesive.

The packs or the online safety sheets of typical adhesives will often list some simple solvents such as acetone, methyl ethyl ketone, ethyl acetate, isopropanol. These are familiar chemicals in nail polish removers and hospital wipes. They can make adhesives much more versatile as the solvents are relatively volatile, allowing a liquid glue to be converted quickly into a "grab" adhesive. Isopropanol is found in "fast cure" plaster fillers – it evaporates faster than the water that is normally used. Their

"Volatile Organic Compound" status and their desirability to glue sniffers means that solvent-based adhesives are increasingly going out of favour.

In addition to the simple solvents you often find some rather complicated chemicals. Here is a short guide to some of them:

- Pentamethylpiperidine sebacate or "Contains Reaction mass of Bis(1,2,2,6,7-pentamthyl-4-piperidyl) sebacate and Methyl 1,2,2,6,7-pentamethyl-4-piperidyl sebacate". These are HALS (Hindered Amine Light Stabilizers) which help sealants and adhesives to survive the UV light outside. Silicones are naturally stable in UV so anything which seeks to replace them for general outdoor use needs to add something like a HALS (along with fillers such as TiO_2 which absorb and scatter UV) to increase the stability.
- (3-Aminopropyl)triethoxysilane, and *N*-(3-(trimethoxysilyl)-propyl)ethylenediamine. These are APTES and a slightly more complex version, as adhesion promoters.
- Dioctyltinbis(acetylacetonate), Dibutyltin dilaurate. These (and just about anything else with "tin" in the name) are catalysts either for silicone and silane reactions or for urethanes.
- Vinyltrimethoxysilane. Small amounts capture low levels of water to keep reactive silane systems stable in the pack. In principle the vinyl part can also take part in some polymerization reactions.
- Hydroquinone. This is a standard antioxidant used in many non-adhesive applications to stop oxygen causing problems with the ingredients. It is specifically a good stabilizer for cyanoacrylate adhesives, helping them to last longer in the tube because oxygen can create radicals that kick off the polymerization. A small amount has little benefit, a large amount makes the adhesive too stable to be cured, so it has to be judged just right to give stability without interfering with reactivity. It is easier to add high levels to those cyanoacrylates that come with activators that readily overcome the inhibition of the hydroquinone.
- 4,4'-Methylenebis(phenyl isocyanate) or anything else with "isocyanate". These are the reactive components of urethane adhesives and need to be labelled as "irritant" and "harmful if inhaled".

- Dibenzoyl Peroxide, or anything else with "peroxide" is likely to be a "radical catalyst". As explained in Chapter 1, these peroxides decompose into radicals that react quickly with a carbon–carbon double bond, creating a carbon radical which reacts with another carbon–carbon double bond and so on. Vinyl polyesters and non-UV acrylates are cured via such radical reactions.

- Things like 1,2-Benzisothiazol-3(2H)-one or "Contains Reaction mass of 6-chloro-2-methyl-4-isothiazolin-3-one and 2-methyl-2Hisothiazol-3-one (3 : 1)" are there as fungicides, to stop your adhesive or sealant going green or black with the sorts of biofilms discussed in the bio-adhesion chapter. The problem is that although we all want our sealants to *not* support the growth of fungi, we also have a fear of "chemicals", especially ones with long names. We want effective biocides which aren't chemicals and which cannot damage any other form of life, which is an impossible demand. Regulators have to juggle the major risks of fungal growths with the minor risks from the fungicides.

5.20   IF ONLY . . .

All of us who use adhesives at home have a simple wish – that we could buy one tube of glue which would: always stay fresh in its tube; never block the nozzle; be easily applied; stay "open" for as long as it takes us to fiddle around till we have brought our pieces together, allowing us to adjust for the mistakes we made when we brought our pieces together. Finally, we want the adhesive to then set instantly to form an indestructible joint, and to work on all possible materials.

Although that list is full of contradictions, modern adhesives could achieve much of what we want if it weren't for a few constraints.

- Small risks from occasional use of chemicals get exaggerated into "we're all going to die" media campaigns so the constraints on what can be used within adhesives get tighter and tighter. It is easy to make a safe glue (flour and water), and not so easy to make a safe and effective glue. It is interesting that superglues haven't come under attack from those who want "chemical free" products. They are, after all, cyanoacrylates – we are using glues containing cyanide groups! Their safety profile happens to be fine; I am just surprised that someone somewhere hasn't created some internet scare story about them. I suspect that neither they, nor the urethanes with their MDI & TDI, would be allowed to enter the market if they were invented today; regulators are ever more cautious and internet scare stories are ever more lurid. Even a mix of flour and water is a problem because microorganisms love to grow in it and a glue full of nasty bacteria can be genuinely toxic to a small child. We can't add an antimicrobial to the flour and water paste because there is always someone who will create an internet scare story that whatever X is added to stop the microorganisms is going to kill us all.

- Because a minority decide that it is a good idea to sniff the otherwise excellent solvents in the adhesive, there are bans on the use of those solvents in adhesives, making it harder to formulate what sensible users would like.
- If users were prepared to spend a few seconds preparing the surface properly (cleaning, sanding ...) and an extra few seconds applying, say, a primer coat, adhesion would be greatly improved. Experience shows that we (this includes myself!) don't like to do the preparation and that we hate having any two-part adhesive system. Epoxies are grudgingly accepted by consumers because they are one of the few systems that cure super-hard; even here we tend to choose the "rapid cure" systems that are generally less good than the more conventional, slower cure.
- A stiff adhesive might be exactly wrong for a joint that needs flexibility. A flexible adhesive might be exactly wrong for a joint that must never yield. A thin adhesive might be perfect for a pure "pull" stress as in a butt joint, while a thicker adhesive might help spread the load for a lap joint.

When we all next complain about our household adhesives, we should remember that their limitations are partly our fault, partly the result of our inability to be rational about small risks, partly because of a minority with a self-destructive urge ... and partly because adhesion is a property of the system and the best adhesive in the world is useless if it is used to do the wrong job in the wrong system.

CHAPTER 6

Strong Adhesion with Weak Polymers

At one time I thought I knew a lot of adhesion science and gave a talk about it at a conference in Stockholm. At the end there were the usual questions which I could handle OK until there was one final question. At the end of the question I had to admit in public that I was so stupid that I didn't even understand the question. I have often found that admitting my ignorance has been the start of a new adventure and this time was no different. In the coffee break I met up with the (as it turned out) famous Russian Professor Mikhail Feldstein, who had asked the question. In approximately 10 minutes he described to me the theory of pressure sensitive adhesives, the common adhesive tapes we all use which give strong adhesion because (surprisingly) they are based on weak polymers. I had been entirely unaware of their science, which is why I was unable to answer his question. When I got home, I immediately downloaded his academic papers and bought the Benedek and Feldstein books which are a standard in the industry. Very quickly I found myself spending more time troubleshooting this sort of weak/strong adhesion than the conventional strong adhesion that had previously been all I knew. My job in this chapter is to translate those 10 minutes of intense Russian theory – and years of further research – into a story that I hope will fascinate you as much as the topic fascinates me.

Sticking Together: The Science of Adhesion
By Steven Abbott
© Steven Abbott 2020
Published by the Royal Society of Chemistry, www.rsc.org

The stars of this chapter are those ordinary tapes we take for granted, often calling them by familiar names ("Scotch tape", "Tesa tape", "Sellotape", "Duck tape", "Band-Aid") even if we are not using (or even aware of) a specific brand. For those who get excited about whether it is duct tape, duck tape or Duck tape, I note the interesting fact that the original word "napron" morphed from "a napron" to "an apron", so maybe duct tape morphed to duck tape or vice versa. Language is never fixed. On the other hand, we can agree that Apollo 13's lunar module air recycling system *was* fixed with the help of duct tape and not with duck or Duck tape. For those who choose to use gaffer tape (rather similar to duct tape but generally easier to peel off), the debate is whether it should be called gaffa tape.

Other than double-sided tapes (where each side is in contact with a release material), all PSA tapes consist of the adhesive on a strong carrier which I call "backing tape" and which others just call "backing". Although the backing tape might at first seem a rather neutral player in the story, we shall see that its properties are a key part of the overall adhesive system.

If you buy a range of adhesive tapes and test them by peeling strips from the digital scales in your kitchen, you will find that there is a large range of adhesive strengths. A sticky note, designed to be easily removable, is around $4\,\mathrm{N\,m^{-1}}$; a typical non-aggressive tape is $100\text{--}200\,\mathrm{N\,m^{-1}}$ and you can find really strong ones (with names of strong animals such as gorillas or dinosaurs) in the $400\text{--}800\,\mathrm{N\,m^{-1}}$ range.

What do these numbers mean? Imagine that I am an action hero and the bad guys are escaping in a helicopter. At the last moment I leap and grab the skid of the helicopter and I'm dangling in the air. The skid is probably 2 m long but for simplicity we'll think of it as the width between my hands, 1 m. I weigh 80 kg which is 800 N. So, when I am dangling from that helicopter the result is $800\,\mathrm{N\,m^{-1}}$, the same as a strong adhesive tape. Switching to a less dramatic scenario, if a typical 20 kg 6 year-old decides to dangle from a 1 m wide bar, that is $200\,\mathrm{N\,m^{-1}}$.

While we are getting a feel for numbers, we also need to think again about that factor known as the "modulus". This was defined in Chapter 2 and I have used various modulus numbers for "strong" adhesives without really making clear what they

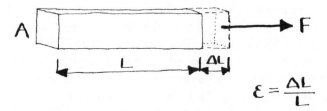

$$\varepsilon = \frac{\Delta L}{L}$$

Figure 6.1 Repeating what we saw in Chapter 2, the modulus is the ratio of the stress (force per unit area) to the strain (the extension ε).

mean. Figure 6.1 is a repeat of the diagram from Chapter 2, and we will now explore it in more detail.

Make a small bar of epoxy with a known width and thickness (the length is not important). If the width is, say, 25 mm and the thickness is 5 mm the cross-sectional area of the bar is 125 mm². Now clamp the bar at each end then start to apply a force, at the same time measuring by how much the bar is stretching under the load. We might find that a load of 1000 N (the weight of a 100 kg person) stretches it by 0.2%, or 0.002. This fractional increase in length is the *strain* on the sample. The force per cross-sectional area (once we have converted it to m²) is $1000/0.000125 = 8$ MN m^{-2}, where the M = mega = million. This is the *stress* on the sample. When we divide this by the strain, we have the *modulus* of the material. In this case $8/0.002 = 4000$ MN m^{-2}. N m^{-2} are also called Pa (Pascal) so the modulus of this epoxy is 4000 MPa, or 4 GPa where G is giga, meaning a billion. This is a typical modulus value for an epoxy.

To repeat some other numbers, the tough PET of your drinks bottle also has a modulus of 4 GPa, a typical PE used for a bag might be 0.5 GPa, with most other polymers somewhere in between. The modulus of aluminium is 69 GPa and that of steel is 200 GPa, similar to that of carbon fibre.

What has this got to do with our adhesive tapes? We shall see later that the modulus of the backing tape makes a big difference to the performance of the tape – PET tapes perform better than LDPE ones even if the adhesive is identical. The key point of interest right now is that you cannot have a general-purpose adhesive tape if the modulus of the adhesive coating is greater than 0.3 MPa. That's right, these adhesives *must* be weak (by a factor of 13 000 compared to an epoxy) in order to be strong. I find that a fascinating fact.

6.1 PSAs – PRESSURE SENSITIVE ADHESIVES

These tapes are typically called PSAs – Pressure Sensitive Adhesives. The name is well-intentioned and deeply wrong. The *intention* was to convey the idea that with a bit of light smoothing pressure to even out any wrinkles when you apply the tape, you have a strong adhesive. The *impression* is that if you stick it down with light pressure you will get a lower adhesion than if you really push down hard. This is simply wrong. Once you've smoothed out any obvious air and wrinkles, the adhesion of a PSA is independent of pressure. And it is all because of the 0.3 MPa.

This number is the famous Dahlquist criterion. In 1969, Dr Carl Dahlquist, working on the relatively new science of PSA at 3 M found, after lots of tests, that: *"For measurable quick tack, the elastic modulus must be below a certain fixed value which is fairly independent of the nature of the adhesive, the adherend and the applied pressure."* That "certain fixed value" is the modulus 0.3 MPa. If you had a PSA formulation with a modulus of 1 MPa it would not naturally stick to most general-purpose surfaces, though it would have no problem sticking to a smooth surface such as glass.

Where does this 0.3 MPa come from? At the start of the book, we discussed how it was possible to get reasonable adhesion if you could bring two surfaces fully together. With super-smooth rubber onto glass, or highly-polished metals, or with a set of tricks on the gecko's feet, this is possible. For everything else it is impossible. As this section is entitled PSA, we can ask the question: "What happens if you *really* press two hard surfaces together – surely you can get perfect contact?" This turns out not to be true (Figure 6.2). You can only get perfect contact by applying infinite pressure.

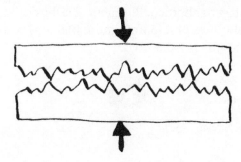

Figure 6.2 No matter how hard you squeeze, you can't force two surfaces together.

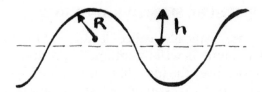

Figure 6.3   This doesn't look like a rough surface, but it turns out that any surface can be characterized as having an average roughness height *h* and a radius of curvature *R*. These two factors then affect how well a PSA will stick to a surface.

At the other extreme, if you have a rough solid and try to get perfect contact with a liquid, well, that's no problem. Indeed, as we've seen for conventional liquid adhesives, we use this trick to get perfect contact first, then ask the adhesive to solidify, becoming, for example, our 4 GPa epoxy.

For a good PSA the first requirement is to achieve that perfect surface energy contact without being too liquid, and this is why PSAs all tend to be near Dahlquist. A PSA with an even lower modulus or, say, 0.01 MPa would be, as we shall see, useless in other ways.

Let us see where Dahlquist comes from. It is an amazing fact of physics that you can take most complicated, rough surfaces and get to their essence with just two numbers: an average roughness height, *h* and an average radius of the rough elements, *R* (Figure 6.3).

If the surface energy is *W* then any material with a modulus less than a critical value (which depends on *W*, *h* and *R*), when put into contact with the peaks of the surface, will spontaneously flow into the valleys and produce perfect contact. For those who are interested, this critical value is $W\sqrt{(R/h^3)}$. If you go to the app (https://www.stevenabbott.co.uk/practical-adhesion/dahlquist.php) and use typical values of *R* (5 μm) and *h* (0.5 μm) you will find that the critical modulus is ~0.3 MPa.

https://youtu.be/7rPvHo_UpC0

If you take a surface to which a tape sticks nicely and roughen it with sandpaper, you will quickly find that the adhesion decreases with increasing roughness. This is seen in the simple video I made a long time ago with a sticky note and smooth and roughened piece of plastic. The reason is that either R gets smaller or h gets larger, so you need a lower modulus adhesive to get perfect contact. The lower modulus to get adhesion onto the rougher surface will generally give weaker adhesion. All PSA tapes face a trade-off between general-purpose strength (higher modulus) and ability to cope with rougher surfaces (lower modulus). This is a reminder that roughening a surface is not a guaranteed way to significantly increase adhesion.

Sometimes we want to make a PSA that is not tacky at room temperature. This means that we can, for example, make a large roll of it without worrying about how to unstick it from itself. How do we then get it to stick? A specific example is the sort of protective seal commonly found under the lids of glass or plastic bottles (or jars). These seals are typically layers of printed paper, aluminium and a non-sticky adhesive, placed into contact with the rim of the bottle. The aluminium needs to be there to provide a near-perfect barrier against oxygen getting in or flavours getting out. It also, very conveniently, absorbs energy rapidly from an inductive electromagnetic field. On the packaging line, the bottle and seal pass under the magnetic field which near-instantly heats the aluminium (via induction) which then melts the adhesive which can then gain perfect contact with the rim of the bottle. Once everything has cooled, we have a strong PSA adhesive. This "induction heat sealing" is done on millions of bottles each day. The impact of these seals on the UK milk bottle industry is discussed in the section on packaging. For those who are interested to know more, I have an app at https://www.stevenabbott.co.uk/abbottapps/HSC-I/index.html.

Is a meltable PSA really a PSA? The question only arises because of the historical accident of naming them as pressure sensitive. These melt-sealed adhesives are still weak (a few MPa) compared to classic strong adhesives, and work in exactly the same manner as the usual PSA tapes.

Perfect, full surface energy contact, via Dahlquist or some form of heat step, is a necessary first step. But remember, though, that surface energy only gives us that feeble 40 mN m^{-1} of peel, which is a long way from the 4 N m^{-1} of a sticky note or the 400 N m^{-1} of a strong tape. What else is required?

6.2 DISSIPATION

In the chapter on strong adhesion, the focus was not so much on the strength of the adhesive as on its ability to absorb the crack energy via dissipation. With an entangled network across the interface, any movement of one part of the network would spread out over a few nm, and the crack energy would get dissipated as heat.

Although they cannot tangle *across* the interface, all PSAs are *self*-tangled; their polymers are not free to slide past each other. Some of them are tangled using classical crosslinks, just like the strong polymers but at a much lower level (a few %) of crosslinks. Others are built from different polymers such as polystyrene and polybutylene ("SBS rubbers", Styrene-Butadiene-Styrene). The polystyrene separates into its own solid domains and is completely useless as a PSA. The polybutylene is very rubbery and on its own is also useless. The trick is that some parts of the polybutylene chains are trapped inside the polystyrene domains which act as points of crosslinking.

To get an idea of how PSAs work, you need some tape, a glass microscope slide and either good close-up eyesight or a magnifying glass or USB microscope (Figure 6.4). Find a way to clamp the slide.

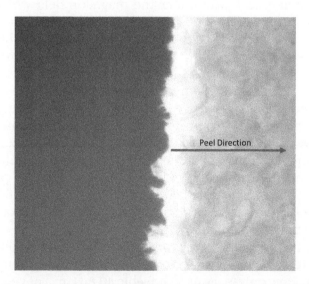

Peel Direction

Figure 6.4 Peeling creates lots of distortion with fingers of adhesive trailing behind. Either they ping off to leave a clean surface (as here) or the fingers split to leave small blobs left behind.

I have used the edge of my desk and a heavy weight to hold one end of the slide, so that the rest of the slide is hanging freely over the edge. Now stick a piece of tape on the underside of the slide and attach a reasonable weight to the tape so it hangs with a significant load on the tape, though not enough for the tape to pull straight off. Although this is relatively easy to do at home, you can also see my video of a few different PSAs being peeled.

https://youtu.be/ALuIjc-el5k

Those "fingers" which are easily visible are bits of the PSA being stretched not by nm but by mm. A diagram (Figure 6.5) of what it looks like from the side makes things clearer, though the compression zone is not visible under most circumstances and is discussed much later:

I hope that the diagram reminds you of the difficulty of removing chewing gum from a surface. Chewing gum has to be a weak polymer (otherwise you could not chew it!) and its stretchiness is part of its appeal. As you try to remove it from a surface the stretchiness is a nuisance – you pull hard and all you get is a string of the stuff, not the whole lump. In the diagram we see that those fingers can do two possible things:

1. They can ping away from the surface and retract back into the bulk of adhesive remaining on the backing tape
2. They can break at a small "neck" in the finger and leave a lump behind on the surface

Figure 6.5 When a PSA is being pulled off, you get fingers still attached to the surface which start to thin, forming a neck. Usually they ping away from the surface, leaving no residue (left image), but sometimes the neck breaks, leaving a small amount of residue behind (right image).

Different tapes are designed to fail in different ways. In most cases we prefer the clean peel, leaving no residue. In others there is a requirement for "cohesive" failure which leaves the residue. One example of this is something that used to annoy me. Often when you open a jar of instant coffee there is a foil seal (created by the induction heating mentioned above) which you have to remove. It always offended me that removal left a residue on the glass; what I wanted was a nice clean glass rim and I was surprised that the manufacturers could not achieve that. I eventually found out that it was a *requirement* of the adhesive to leave the residue! This is because with a clean-break system the user would have no way of telling whether there had been a defect in the seal, with no effective adhesion and, therefore, the prospective loss of flavour or ingress of contamination.

When you see the cohesive failure, it is easy to imagine that the fingers broke halfway across. With our glass slide and tape, we can readily show that this is not, in general, the case.

Get a hair drier and heat the glass from the top side while gently testing the tape. With most common tapes, as the adhesive gets hotter the polymer flows more easily and also snaps more easily. As you peel, you see a residue left on the slide (Figure 6.6).

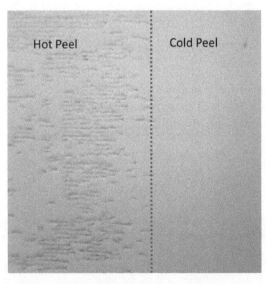

Figure 6.6 A household tape leaves a (small amount of) residue when peeled hot; peeled cold there is no residue.

If you then look on the tape itself you see that it is almost the same as the cold peel portion, i.e. the residue left on the slide is a small fraction, showing that the fingers snapped near the slide rather than in the middle of the finger.

How does the PSA decide whether to ping off or snap? When I was writing my Adhesion Science book I was greatly bothered by this question and felt very stupid because I could not find any academic papers that that could help me explain the rules to myself so that I could explain them to my readers. Fortunately, at that time I went to a high-powered adhesion science conference and met one of the true great names of PSA science, Prof. Constantino Creton. I explained that I felt stupid not being able to predict this failure and he laughed. "We know in principle how to calculate this, but the calculations are too complex, covering too many time- and length-scales. This makes it impossible in practice". It greatly cheered me up that my inability to calculate how the system split wasn't my personal failure.

I later found a wonderful piece of research that probed this splitting further. Using a complex setup which made it possible to peel the system at different speeds and at different temperatures it was possible to see that the transition from clean peel to cohesive failure went via a noisy "stick-slip" regime where the tape oscillated between failure modes. Readers who remember an earlier chapter explaining that temperature and time are equivalent will not be surprised that this research showed that you could get the stick-slip at one speed at one temperature and also with a different speed at a different temperature. Using a standard conversion trick between time and temperature (WLF, as briefly mentioned in Chapter 4), they could construct a whole curve of the PSA's adhesion, going from very low at very low speeds (the adhesive had time to separate with little effort) through a maximum then through stick-slip and finally to the sort of low adhesion that, as nurses know with plasters, comes with very high speeds.

The science says that high speeds are equivalent to low temperatures. Having done the experiment with a hair drier, if you have some ice, dry ice or liquid nitrogen (or, as described in Chapter 10, a squirt from a clean air spray can) to cool your glass slide, you can get a good idea of how your adhesion changes at lower temperature. When really cold the adhesion will be pure surface energy (as there is no possibility of dissipation in the rigid matrix) and therefore the tape will pull off very easily. This all means that the curve of peel from low speed to high can be described equally in terms of temperature from high to low. The key point is that you cannot say, "This PSA gives cohesive failure and that one gives clean peel" because each of them, at some speed or temperature will show either behaviour.

One day I was helping a company sort out some problems with their PSAs and they complained that they could not properly check the adhesive on its own because it was stuck to the backing tape. I simply said "Liquid nitrogen" and they looked at me greatly puzzled. They didn't have any on site but were friendly with a local university who had lots of it. A short time later they came back looking very happy. After putting the tape into the liquid nitrogen, it was trivial to pull the now-rigid adhesive off its backing tape. We were then able to measure its properties on its own (at room temperature) and the results turned out to be very useful.

I was a bit lucky with this because if you try this trick with a normal adhesive tape it often does not work. Which is a good thing in many ways. We want the adhesive to be easy to apply and remove from the surface, while always remaining stuck to the backing tape. It would be very annoying if you pulled the backing tape and left a layer of adhesive behind. Fortunately, we (and the tape manufacturers) already know how to get strong adhesion of the adhesive to the backing tape – use whatever trick creates entanglement across the boundary. The difference between entangled adhesion and PSA adhesion is so large that you will rarely get failure of the adhesive at the backing tape interface.

There is one more thing to say in terms of dissipation. Any energy which is produced by the tape is a welcome diversion away from the crack propagation. The main dissipation is the heat produced as the polymer slides over itself when being stretched. There are two more forms of energy that are emitted. The first is sound – there is always some noise produced by vibrations in the tape being transferred to the air. The second is in the form of high energy electromagnetic radiation from X-rays through UV to visible light. When scientists meet a phenomenon that has a cause (in this case, rubbing which, in Greek gives us tribo-) and effect (in this case, emission of electromagnetic radiation, which in

Latin gives us -luminescence) but no explanation we tend to say "The emissions are triboluminescence" as if that explains something, rather than restating the effect in fancy mixed language. Eventually we find an explanation and then the word describes the effect and is shorthand for the cause.

With triboluminescence we know that violently ripping surfaces apart will generate charges on opposite sides of the break, though exactly how is unclear. We also know from lightning that if you have large voltage across a gap then it can break down the air in a more or less spectacular manner, emitting light. That's the cause of the visible triboluminescence. The breakdown in air cannot generate X-rays (so tapes are safe to use). To obtain the X-rays (via bremsstrahlung of free electrons) the tape has to be pulled apart in a partial vacuum. This then leads us to questions about bremsstrahlung but we have to stop the process somewhere, so maybe we can just agree that the blue light and X-rays are caused by triboluminescence.

Returning to the sound emission, I was once giving a talk along the lines of "throw your adhesion problems at me and I'll try to solve them", which is always exciting. Someone at the back started to ask a complicated question about adhesion of a label to a yoghurt pot and proceeded to peel the label from a pot he'd bought to the seminar. The sound was rather staccato, not at all like a smooth PSA peel. With total apparent confidence, while still walking over to examine the pot, I said; "Oh, the adhesive is far too rigid, you need to modify it with blah blah. . ." It turns out that my instant diagnosis was correct. It might easily have been wrong, but that's how science works. You propose a root cause diagnosis and cure; if the evidence supports it, then there's a clear action and everyone's happy. If there is extra evidence which contradicts it, there's no shame in being wrong – by eliminating one possibility you've already made progress towards the real cause.

6.3   SPAGHETTI (COOKED) AND CHEWING GUM

Many years ago, spaghetti was still somewhat exotic (hard to believe!) and knowing when it was properly cooked was a problem. I don't know why anyone would think of such an idea, but I was told that to know if it was cooked you should throw a piece of spaghetti at the ceiling. If it fell straight off, it was undercooked, if it stuck somewhat, then fell off, it was just right and if it stuck hard on the ceiling then (a) it was overcooked and (b) you would have a problem removing it.

What I did not appreciate at the time is that this beautifully shows a number of key aspects of PSA. When they are too hard (undercooked) they will not stick. When they are the right degree of softness then they can stick – and will become unstuck if they become too hard, which is what happens when the cooked spaghetti cools and falls off the ceiling. When they are over-soft, they can still stick but will be messy and hard to remove cleanly – which is the overcooked spaghetti coated with soft starch.

The spaghetti story captures the essence of something that everyone knows when they touch the surface of a tape, yet is hard to pin down. The scientific word for this feeling of "stickiness" is tack.

Chewing gum shows a similar set of property variations. As a "stick of gum" it is not at all sticky, though bits of powder on its surface help avoid stickiness if the gum absorbs moisture within its wrapping. When it is chewed, the water softens the gum, making it stickier. The rubber polymer which is the basis of chewing gum is unaffected by water, but it makes up only 30% of the original, with the other 70% being more able to absorb water.

Those of us who don't live in Singapore (where chewing gum is generally banned) have to deal with those who throw their lumps of chewing gum onto the street once they have lost their flavour. If it is stepped on or driven over by a car, it is soft enough to make good contact with the surface. If it is a hot day, trying to pick up the moist, hot lump will be an unpleasant experience. After some days of drying out as a squashed layer, and if rain has washed out any remaining hydrophilic components then we have a lump of sticky rubber acting as a PSA. This is especially hard to remove because it is thin (difficult to get underneath to peel it) and a dissipative rubber. The rubber itself only breaks down slowly because it is a hydrocarbon with no links that are easy to attack in the environment. As we saw in Chapter 2, we still have birch bark pine gum (another hydrocarbon) with teeth marks (and DNA) from 10000 + years ago.

Although we are unlikely to have chewing gum police using DNA to track down those who litter the streets, it is gratifying to know that criminals who grew tired of their chewing gum while committing a crime have been tracked down via the DNA in their discarded gum.

6.4 TACK

Everyone knows the difference between a "tacky" and a "non-tacky" PSA. You just touch the adhesive and form an intuitive judgement. A tacky tape gives an instant grab and it feels as though there is some tendency to create strings/fingers of adhesive as you pull away from the surface. It is therefore frustrating that there is little agreement about what tack actually is and how to measure it. Tack is not the same as peel strength and it certainly isn't shear strength. So two of the standard tests for PSA don't relate in any direct way to tack. While too little tack makes it unlikely that the adhesive will stick (because it fails Dahlquist), many apparently tacky/grabby adhesives are rather too liquid to provide adequate adhesion when you need it. This is especially true if you need significant shear resistance. Later, we will discuss the "probe tack test" which, from its name, would be expected to be a direct measure of tack. But just like adhesion, tack is a property of the system. The probe tack test doesn't give us any direct, reliable measure of tack or of anything of direct benefit to the design of an optimal PSA tape.

6.5 THE BACKING TAPE

In Chapter 4 on measurement we saw this app image (www.stevenabbott.co.uk/practical-adhesion/peel.php) showing how the peel forces are spread ahead of the peel separation point (Figure 6.7).

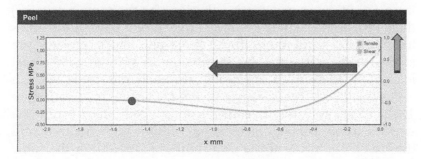

Figure 6.7 The effective peel strength depends on how the stresses are distributed ahead of the peel front.

The way these forces are spread out is important; if the stresses are concentrated in the first fraction of a millimetre, the chances are that the bond will fail because there is little chance of general dissipation ahead of the peel. The way those forces concentrate depend, of course, on the adhesive's modulus and thickness. Because designing for the Dahlquist criterion leaves us little room to change the modulus of the adhesive, this leaves us with the thickness of the adhesive layer.

While reading the claims of a super-tough tape I was interested to see that the adhesive was "twice the thickness". I happened to know that the basic theory of a PSA (it is in the equation Prof. Feldstein showed me over coffee) says that peel strength is proportional to thickness. I also knew that although the experimental evidence for this is limited, everyone agrees that a very thin PSA layer is unsatisfactory. Looking at the physics from the app, what stands out is that doubling the thickness can very much spread the load ahead of the peel zone. This makes the "twice the thickness" claim plausible. As a bonus, it means that on rough surfaces there is enough thickness of adhesive to flow into good contact.

https://youtu.be/7yfWbzpRc_M

These tough tapes come with a tough, reinforced backing. Conventional tapes use a plain polymer backing tape. If this is weak (low modulus) it automatically concentrates the stresses near the peel zone, which is equivalent to having a thin adhesive layer. So, PE tapes give lower adhesion (for the same thickness) than PET-backed tapes because PET's modulus is 6–10 times higher. The reinforcing fibres in strong tapes provide the equivalent of extra modulus in the tape, ensuring that the peel load is spread as much as possible. If you are interested, you can check this in the app. Change the modulus from 0.5 GPa (PE) to 4 GPa (PET) to 10 GPa (equivalent to a reinforced tape); the stresses are spread out much further with the high modulus, and the maximum stress at the peel front is reduced. For the same adhesive, the effective strength of the tape will be larger for the higher modulus backing tapes. If you *have* to use a low-modulus backing tape then (as you can check with the app), increasing its thickness is equivalent to increasing its modulus.

Which brings us to duct tape which is fabric based. The tape itself is not especially strong – indeed it is designed to be easy to rip across the width so you can cut off a piece without scissors. This ease of breaking is apparent (spoiler alert!) in the hostage video discussed shortly. The relevant strength is with the fabric fibres in the peel direction, ensuring the same broad peel zone as a fibre-reinforced PET tape.

Now, although the reinforcement makes the general tape tougher, I don't think that this is the main reason for its existence – normal tape backings are strong enough for many purposes. The main reason is because it solves a different problem.

There is a huge issue when making super-strong PSA tapes. It is the opposite to the well-known Teflon pan problem discussed later: "If nothing sticks to Teflon, how does Teflon stick to the pan?". In this case the problem is "If the tape sticks super-well to any surface, how can we make a tape that can be unwound from its own roll?" One answer is to provide a silicone release layer to the reverse side of the tape. A more elegant answer (because adding silicone release layers is neither easy nor cheap) is to make the back of the tape full of crack defects, i.e. sites where the adhesive has no chance of sticking. Usually we avoid sites with poor adhesion because cracks easily start at these points (remember Griffith's law from Chapter 2) and once they get going, they are hard to stop. This normally negative behaviour is exactly what we need to solve our strong PSA tape problem.

The type of strong tape with reinforcement fibres is naturally prone to creating cracks simply because the surface is so uneven. When I first examined a pack of "animal strength" tape, which was not fibre-based, I was distressed to see a rather glossy surface, which did not seem to be rough. When I gave an experimental tug to un-wind a bit of the tape, I was alarmed that it needed such a large force – then found that it carried on unwinding easily, with a jerky motion that was unlike anything I had previous encountered. With a low-powered microscope (or the good camera lens used in the video) to look more closely at the glossy back surface, the explanation for the behaviours became clear. The surface has a regular pattern of rather deep indents that the adhesive cannot flow into. This means that the adhesion from the adhesive to the next layer of backing material is set up for failure, allowing this genuinely strong adhesive to be removed rather easily. Well, not quite. With any new roll of tape, I test the adhesion at as slow a rate as I can manage, then as fast as I can manage. With this tape there is a big difference between the two peel rates. At a slow rate we have to get the adhesion to fail at each new indent, as well as pulling it apart in the flat zones between the indents – on average the adhesion is rather strong. If we try it quickly, the failure at one indent rushes along the interface to the next indent, so the cracks continue propagating, making the adhesion remarkably small. How can a failure "rush along"? Because time is equivalent to temperature. At high speed, the adhesive responds as if it is very cold, with a much lower adhesion.

Out of curiosity, I tested the hypothesis. I first stuck the tape to a glass slide, at room temperature, to let me test the downside of the release mechanism. Sure enough, I could see through the slide a replica of the pattern of the backing tape. The adhesive surface had partly flowed into the backing indents and so its surface was rough. This pattern meant that the still-rough tape was only half stuck to the glass. With some effort I could squeeze the adhesive into much better contact. In this case the tape really is pressure sensitive, though at room temperature it required considerable pressure to gain good contact. When I warmed the slide, it was much easier to get the flow required for good contact, and the adhesion was indeed large. Then I put the strongly adhering tape into my freezer. After 30 min I took it out and could easily peel off the tape – cold tape really is equivalent to a fast peel, just as the theory says it should be.

6.6   HOW DOES A PSA WORK?

The short answer is "it's complicated". Because I cannot offer a neat explanation (none exists), my aim here is to give you an appreciation of why your tape might be too sticky or not sticky enough, why it leaves a residue or doesn't leave one, why it is too noisy or too quiet. If the knowledge helps you choose a better tape for any given application, that's a bonus. Before offering my best explanation, we need to better understand why there isn't a simple one.

We already know that we cannot say "The peel strength of this tape must be X"; this is because X depends on temperature and timescale. Many researchers got lost in the complications of temperature and time, until Dr EP Chang at Avery Dennison defined his famous "window". He said that:

1. A typical adhesive tape has to cope with timescales of 0.01 to 100 seconds. The 0.01 s is a typical timescale for peel. The 100 s is a typical timescale for the tape slipping in shear – e.g. trying to slide an already stuck label along the surface.
2. All adhesives have a measurable "elastic" modulus, G' and "loss" modulus, G'', which respectively describe how the material responds with recovery (elastic) or flow (loss) under stresses applied over different timescales. He then noted that PSAs worked in a window of G' and G'' values defined at test rates of 0.01 and 100 per second. Those who want to see the Chang window in more detail can visit the app at https://www.stevenabbott.co.uk/practical-adhesion/chang.php. The point is that a PSA must not be too weak in terms of conventional modulus (G'), nor may it be too strong – otherwise it exceeds the Dahlquist criterion. At the same time, it *must* be able to flow like a liquid (G'') – not too much and not too little.
3. There are a number of potential windows depending on the temperature range required for the PSA. For example, labels for items in your freezer must be different from those that have to stay on a tractor working all day in hot fields. Each would fit in the standard Chang window *if they were measured at their working temperature*. Because of temperature/ time equivalence, these different types of labels appear at different timescales and values because the window is based on measurements at room temperature.

As has often been noted, any formulation outside the chosen Chang window is unlikely to work, though plenty of formulations inside the window might still fail. Which means that other factors are important.

How, then, does a formulator achieve the right balance of properties? A specific example is the common household tapes based on acrylic PSAs. At their adhesive heart is a polymer like PMMA, the familiar Perspex or Plexiglas. Clearly PMMA is too rigid; the regular 1-carbon methyl groups (the first M) can pack to create a classically rigid polymer. An alternative would be PHMA with Hexyl, 6-carbon, groups. This is just a weak and useless polymer. A blend of methyl (1-), butyl (4-) and hexyl (6-) gets us a long way towards the required curious mixture of strength and flow.

The problem is that once this polymer starts flowing, there's nothing to stop it. Such an adhesive would be nice and sticky but not very strong. A small percentage of a 2-functional acrylate can be added to create crosslinks between polymer chains. As the polymer chains are stretched, they flow nicely past each other until they reach the limits of the crosslinks which bring the flow to a halt. This aspect of a PSA is crucial (it is hard to make a PSA without crosslinks) but difficult to pin down. Adding more crosslinks will help make a tougher PSA until suddenly it ceases to be a PSA because it is too hard. The crosslinks are there to give us entanglement, because it is the tangles that allow the system to soak up the crack energy.

How much entanglement is required? It must be enough to give us the Chang window balance of G' and G''. But those tests involve very small strains which are nothing like the ones that show up as fingers in the peeling tape. To look at those we need a tensile test that stretches the PSA to a much larger extent.

We can imagine our tensile tester measuring four different formulations. The graph (Figure 6.8) shows the force per unit area (stress) against the fractional extension (strain) for these materials. Number 1 might be a classic strong polymer or an over-crosslinked PSA. You need a lot of stress to achieve a rather small strain, and then the whole system breaks. At the other extreme is number 4 which is a soft polymer with no crosslinks. It gives very large strains for rather little stress and is equivalent to trying to make a PSA using treacle or molasses – definitely sticky but not very effective. Something like number 2 seems to be ideal for a PSA. It is reasonably strong at the start, yet when it yields, the crosslinks provide plenty of resistance so that the stress remains high while the strain increases – i.e. the fingers grow as the tape is pulled, as we saw in the pictures of the PSA being peeled from a glass slide. Finally, those crosslinks reach their limit and snap.

As mentioned, the problem for Chang and others is that the G' and G'' measurements are made with very small strains, maybe within the little domain shown in the figure by the circle. For these large-strain tensile tests we need big samples of the PSA. When I referred to the liquid nitrogen trick for getting hold of the PSA itself, it was to obtain these tensile measurements. The

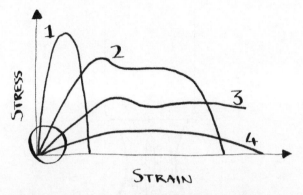

Figure 6.8 A tensile test records the stress (pull) needed to create a given strain (stretching). Here we have strong adhesives (1) that are useless as PSAs, sticky/flowy materials (4) that are useless in a different way. Then we have (2) as a good PSA behaviour, with (3) being what you might find if (2) was measured at higher temperatures or slower speeds. The circle shows the small area of stress & strain covered by standard G' techniques.

results were worth the effort because samples that were rather similar when measured in the small G' and G'' window showed big differences in the tensile tests – differences that could be related back to the degree of crosslinking.

These tensile tests are not so common in the PSA world. A popular alternative test is a "probe tack tester" which squashes a sample underneath a circular probe then measures the forces as the probe is pulled vertically. With video cameras looking from the side and from below, all sorts of things can be found such as stringing and cavitation (Figure 6.9). Depending (of course) on the temperature and speed and, less obviously, on the thickness of the sample, the forces and visual phenomena vary over a wide range.

Two things can be said about the probe tack test:

1. All expert PSA formulators use it and spend hours analysing the data
2. No one knows how to relate the data from the probe tack test directly to a successful formulation

There is yet another test that PSA developers have to carry out because PSAs have to get many things right within one

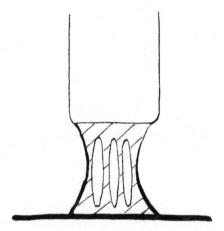

Figure 6.9 In the Probe Tack Test, a layer of adhesive is pulled up by a solid steel probe and starts to cavitate (holes appear inside) and form fingers.

formulation. We might have a PSA that passes all relevant peel tests yet, because the item in use is exposed to shear forces, it fails because it slides out of position. There is a standard test for this: a sample of known length and width is stuck to a vertical wall, a known weight is applied, and the time taken for the sample to fall off the wall is the "shear resistance" (Figure 6.10).

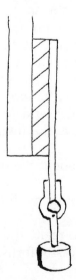

Figure 6.10 The shear resistance test hangs a weight from the tape and the time taken for the tape to slide (shear) down and fall off is measured.

There is a scientific formula for this and, because this is adhesion science, the formula is surprisingly useless. Some careful research showed why. Most reasonable PSAs failed via the adhesive *peeling* away from the wall rather than via the shear the test was supposed to be measuring. This is frustrating for the formulator because, as far as I know, there is no way to scientifically tune the easily measured properties (such as G' and G'') in a way that would predict how the formulation would be able to pass whatever shear resistance specification is demanded by the customer.

For those who like easy-to-remove PSA hooks, this mixing of shear and peel effects is very real, as we shall see later in this chapter.

Now we know how and why PSAs are complicated; how, then, do they work? It is a four-stage process:

1. They establish immediate good contact with the surface, naturally flowing to fill all the minor roughnesses. This happens only if they are soft enough to meet the Dahlquist criterion. Any extra contact from wiping or pressure is to overcome major problems such as bubbles or defects from the tape itself. Those defects might be from roughness of the backing tape or from the adhesive fingering generated when the tape is peeled from the backing.

2. They cling on to the surface with just the tiny ($\sim$40 mN m^{-1}) surface energy.

3. The energy from the peel pull that should create a crack along the interface is, instead, dissipated by producing the long fingers. This is the PSA adhesion that we measure. The weak polymer is strong enough to put up plenty of resistance (so it must be crosslinked) while sliding past itself enough to dissipate the energy.

4. Finally, the fingers are stretched so much that the surface energy forces are overcome and the fingers ping off, leaving a clean surface. Or, depending on the formulation, speed and temperature, the fingers snap near the surface, leaving small blobs of adhesive behind.

The frustration about PSA design is that the four-part explanation is intuitively correct while being hard to translate into relevant measurements. We need the whole package of peel, tensile, G' and G'', probe tack and shear experiments to get access to the big picture, with no known rules about which combination of properties guarantee success.

The "long fingers" explanation is only partially true. Nobel prize winner de Gennes proposed his alternative "trumpet" theory (described in my Adhesion Science book) which is a rather elegant way to think through what is happening in a PSA. The theory tells us that the visible fingers are small players in the overall story, with most of the excitement happening much earlier, on the μm rather than the mm scale. Various attempts to measure how much energy is absorbed at which timescales and dimension scales are ambiguous. My personal view is that de Gennes' ideas are more important than those around the fingering. My key reason for believing this stems from the silicone release science described in Chapter 7. For most of us, most of the time, the finger explanation is good enough because it emphasizes the dual issues of flow (to create the finger) and resistance (fingers of treacle aren't effective), which allow us to think of the polymer itself being a mixture of easy-flow and entangled behaviours.

6.7   STICKING TO DIFFERENT SURFACES

Going back to Dahlquist, we know that if a surface isn't too rough then a PSA is basically going to stick. Although surface energy is necessary (you could not get adhesion to a surface with zero surface energy, though no such surfaces exist), far more important is the amplification by the PSA. This means that you can get strong adhesion to a low surface energy surface such as PE as well as to medium surface energy polymers such as PMMA and to steel, aluminium and glass which have medium to high surface energies. The "strong tape" video shows that you can easily lift up a "non-stick" frying pan because the tape has no problem sticking to the low-energy Teflon surface. You can equally tune the PSA to have low adhesion to all such surfaces – as in the sticky notes we all use.

https://youtu.be/7yfWbzpRc_M

Although PSAs stick to anything, it is still the case that for any given PSA, the peel measured on different surfaces will be different. Via an amazingly lazy and unthoughtful set of citations of citations, the myth is that "PSA adhesion depends on surface energy". It is straightforward to show that this is false. What was at the time the Dow company kindly sent me a high-quality set of peel test values of two different adhesives tested on six different substrates. While it is true that adhesion to PE was lower than adhesion to, say, PMMA, any statistician looking at the data would laugh at a claim that there was a correlation with surface energy. The data are in my Adhesion Science book for those who want a closer look.

Given that the adhesion definitely changes between surfaces, what *does* it depend on? My personal view is that it depends on how relaxed (or not) the PSA polymer is next to the surface. If it is relaxed, it takes a lot of energy to disturb it. If it is stressed, then it is much easier for a crack to disturb it. The degree of relaxation depends on how the molecules at the respective surfaces interact with each other. This is much more complex than "surface energy", which is why there are, so far, no obvious strong correlations between types of PSA and their relative peel strengths on different surfaces.

This explanation slots in as part of the explanation for another puzzle. It is often observed that peel strengths increase over time. The lazy explanation is that this is because the PSA can better flow into small cracks and/or displace air bubbles. That is certainly part of the explanation on some surfaces because things like trapped air bubbles can be directly observed. Yet there are plenty of cases where there is no evidence for such effects. My view is that the PSA polymers have time to become more relaxed, and that increases the adhesion. Why do they need time? Because the polymer chains have been ripped from the backing tape and suddenly placed into contact with the surface, giving them no time to find a relaxed conformation.

You would think that the cause of surface-dependent adhesion could be easily resolved. Sadly, the majority of work on PSAs has been done with bad, or at least thoughtless, science. In my discussions with the relatively few PSA scientists who are probing the correct science, their view is that although they agree with the sentiment behind the "relaxed PSAs stick better" idea, they would express it in rather different language, which would allow them to test their ideas rigorously. If they can provide a convincing theoretical language and experimental proof, that will be a major breakthrough for the field.

6.8   CHOOSING THE RIGHT PSA FOR YOUR JOB

Even though it is hard to pin down exactly how a PSA works, we can still use the general ideas to help us choose the right PSA for a job.

For an easy-peel PSA the main trick is for it to have a modulus slightly above Dahlquist's 0.3 MPa so that it can never fully flow into contact. How would you know this? A test of tack with your finger, although imprecise, is a good start. In the writing of this book I have been amazed at how much can be learned by sticking a piece of tape onto a glass slide and looking at the contact through the glass. A specific example is masking tape which must not stick too well, even if left on the surface for some time. For easy peel from its own roll, the backing must be rough. By sticking a sample to a glass slide and looking through the glass you see that the rough adhesive, even over a few days, does not flow out completely – confirming that the modulus is borderline Dahlquist. The system remains pre-cracked and able to fail relatively easily. Of course, a masking tape that falls off too easily is not appreciated, so tuning the roughness and modulus is a balancing act, made especially difficult by the constant change of paint formulations driven (as discussed in Chapter 8) by environmental concerns.

For a super-strong tape, the backing itself must be strong. Because adhesion to the rear of the backing tape would make it impossible to peel a length of tape, there has to be plenty of roughness on the back of the tape itself, though the pattern this makes on the adhesive will, sadly, reduce the initial adhesion.

Although many tapes advertise themselves as doing every-thing, they cannot fight the laws of physics. At a cold-enough temperature all PSAs become more rigid and therefore lose their ability to absorb crack energy. At a high-enough temperature, all PSAs become softer and more easily stretched, therefore losing their ability to absorb the crack energy. Any tape on a dirty/oily surface will have low adhesion because it will stick to the dirt/oil and not to the surface. Very rough surfaces (such as bricks or cement) are impossible for a conventional PSA to stick to because the there is no chance that the adhesive can flow into significant contact. Even if a special tape can flow into the brick roughness, if the surface is full of dust from construction, the tape will not be able to hold.

In fairness to the tape manufacturers, although the large-scale print on their packs imply that the tapes will stick to anything, the fine print warns against dirt, oil, water and extremes of temperature. One pack had a helpful thermometer symbol saying that the tape could work from −5 to 60 °C. This would be fine for my UK location, but hopeless for those living in colder climates. There are, therefore, special cold-weather tapes (some go down to −40 °C) which will (I expect) be hopeless at normal temperatures, and high temperature tapes that will be hopeless at cold temperatures. The reason for the limitations on temperature ranges is that for many polymers there is a natural domain of relatively flat performance versus temperature and this might cover a range of 50–60 °C. By adjusting the polymer and the additives such as tackifiers used in PSAs, the centre of that range might be placed at 0, 20 or 40 °C. At the bottom end of the range the polymer chains find it hard to move around and the polymer becomes too rigid. At the top end of the range, the polymer chains can move so easily that the system is now just a runny liquid rather than an adhesive.

It is relatively easy to make a water-resistant tape because just about any polymeric or fabric construction will be reasonably water resistant and most of the standard adhesives are not especially affected by water. To declare a tape fully waterproof is much harder because the surface (rather than the tape) might attract water to the interface, destroying adhesion. Once there is water at the interface, either at the start (a wet surface) or over time (water getting to the interface), the PSA ends up sticking to the water, which means no adhesion. Some tapes (and their adverts) imply (but don't claim!) that they can stick to a wet surface. This is impossible unless:

- They can provide tyre-tread features that let the water escape. Treads are necessary because a thin layer of water will not magically disappear into the tape and, as discussed earlier,

Stefan's law doesn't allow the complete squeeze-out of a liquid between two smooth surfaces. An example is tapes used for repairing swimming pools under water. Even these don't work if you just push them into contact; they must be applied with a squeegee to give sufficient local pressure to squeeze the water through the tyre tread cracks. In Chapter 11 we see how tree frogs use the same trick to sweep away the water and achieve normal gecko-style adhesion.

- The adhesive itself is activated by water (i.e. becomes sufficiently low modulus to stick) as with some denture and/or stoma adhesives.

So what about those adverts where a tape is able to seal water gushing out of a large hole in the side of a tank? I tested this in the video on the strong tape. Having confirmed that a typical strong tape had zero adhesion to a wet surface, I tried it on a small-scale replica of the advert. It was no surprise that it worked fine, because the splash of water onto the back of the tape and onto the side of the tank was not enough to prevent most of the tape remaining dry and sticking to the dry portions of the tank, giving an adequate seal.

When you have a dirty/oily/wet surface that needs to be sealed with tape (e.g. to stop a leak) then the only way to do it is via a non-PSA tape. As mentioned in Chapter 3, these have various names such as "F4" or "self-amalgamating" or "self-fusing" and they don't care about the surface as they already have zero adhesion to it. Instead the soft, smooth tape is wound around the surface then in layers around itself, sticking to itself via a gecko-plus adhesion, with the plus part being the fusion of the layers over time. This means that the tapes are formulated so they can flow together over time, which in turn means that the tapes cannot be sold as simple rolls. Instead they have a release liner between each layer.

Professionals who need tape to work for long periods in wet conditions (for example on roof fixings) know that there are some key rules:

- Dry the surface before sticking the tape to it.
- Make sure that the tape joint is *not* on a horizontal surface where water can pool up and gradually undermine the adhesion. Another way to say this is "Never have a flat roof".

- A tape that is optimal for one surface is not necessarily optimal for another. This is frustrating advice because it is not obvious how to choose between different types of tapes for long-term use.
- Choose carefully between thick and thin backing tapes:
  - Thick tapes can be stronger in peel because of the physics of backing tapes – but usually in these jobs peel strength is not important
  - Thick tapes can also have plenty of resistance to UV light
  - Thin tapes can more easily adapt themselves to uneven surfaces, odd curves *etc.* without imposing their own stresses on the joint
  - And thin tapes, edge on, don't provide such a barrier to water flowing over and away from the tape

What about fixing ducts? The official answer, in a report from Lawrence Berkeley National Lab (and in many building codes) is that duct tape is *not* to be used because the fabric/rubber combination is not up to the job. The report mentioned that just about any other reasonable tape (such as a packing tape) did a better job, though officially recommended specialist tapes are the ones to use. If your duct (or, in the case of Apollo 13, your carbon dioxide removal canister) is in a damaged spaceship and duct tape is the only option and has to last only for a few days then, of course, feel free to use it.

In the interests of science, I had to check whether indeed it was a myth that duct tape was great for keeping hostages quiet. When I stuck a large piece of duct tape over and around my mouth (don't try this at home, I'm a highly-trained scientist) I found:

- It came off my bottom lip in less than a second, allowing me to talk normally.
- The removal of the rest of the tape was remarkably easy and painless.

https://youtu.be/iXjklgrQms8

Rather than ask you to take my word for it, and because I was rather ashamed that I had not had the courage to wrap the tape around my head, I arranged for my granddaughter to bind my legs, arms and face with duct tape. From the video you can see that the myth is even less convincing.

It is no great surprise that the adhesion of duct tape around the mouth is poor. The lips are wet and we know that tapes can't stick to wet surfaces. The lips are movable so we can apply strange shear and peel forces that loosen the tape. Also the face (or my face anyway) is rather too rough for good adhesion.

It was also trivial to snap the tape around my legs and arms (after all, duct tape is *designed* to be easily torn across its width) and undoing it from around my head was entirely painless. In any case, duct tape must have poor adhesion to the rough surfaces of clothes or skin because the roughness is outside the limit of Dahlquist flow.

Before leaving duct tape I have to comment on an adhesive challenge of which I had not been aware before researching this book – making a prom dress or tuxedo from rolls of multi-coloured tape. Having seen some award-winning costumes (yes, there are duct tape creative competitions) I can admit to the skill, creativity and hard work behind them while keeping a strong preference for more conventional fabrics.

What about adhesion to *specific* surfaces? The beauty of PSAs is that they don't care all that much about the surface onto which they are sticking. Most tapes will work OK on most surfaces. Stainless steel is a standard test surface and so most PSAs stick well to it by design. Nevertheless, the science we learned about strong, conventional adhesion is relevant here. Acrylic-based PSAs stick especially well to acrylic surfaces such as PMMA – the polymers at the interface are happy and relaxed with low stresses. PE and PP show generally poor PSA adhesion and acrylics do especially badly on them. We have already discussed the hypothesis that the adhesive polymers next to an unfamiliar surface are highly strained and less able to absorb the crack energy. Rubber adhesives, which are mostly hydrocarbons, stick relatively less badly to the hydro-carbon PE and PP surfaces and may, over time, create a strong bond as the polymer chains slowly intermingle and entangle. For adhesion to humans via plasters or stoma tapes, the ability to deal with moisture is important. Many such adhesives contain "hydrocolloids", a fancy word for things like starch which not only absorb water but gain PSA strength by creating the sort of sticky structure that holds over-cooked spaghetti together.

The other type of surface specificity is roughness. As discussed earlier, a thicker coating of a somewhat lower modulus adhesive increases the chance of sufficient flow into a rougher structure, though this trick has its natural limits.

Then we have to think beyond the surface, about the environment. Is the job subjected to lots of shear (so a less flowable adhesive is desirable), or is it subjected to lots of short, sharp shocks, in which case we need the equivalent of a cold weather adhesive because high speed is the same as low temperature?

The last point is deciding whether we care, or not, about leaving a residue when the tape is removed. People often debate which is better: gaffer tape or duct tape. Each is strong yet easy to cut by ripping the tape. The gaffer tape is a "stronger" adhesive in that its modulus is higher, which means that *if* it can flow into contact with the surface it will have strong adhesion, yet when pulled, the adhesive will cohere and come off with no residue. The duct tape is a "weaker" adhesive in that its modulus is lower and might leave a residue on removal. Yet its adhesion might be stronger because it can better flow into the whole structure and stick with fewer cracks. The reason the debate about gaffer and duct tape will never be resolved is that you can get duct tape to perform like gaffer tape, and vice versa, depending on the surface roughness, temperature and time.

6.9 THOSE EASY-REMOVE HOOKS

https://youtu.be/xtGOhL5UVVE

In Chapter 3 we saw that pure surface energy, gecko-style adhesives can stick adequately to smooth glass, yet are easy to remove by peeling when required. Similarly, if a sudden shock hits them then, like a gecko's foot, or like Spider-Man in my video, they easily detach. What about the PSA-style hooks that are removed by pulling on a tab below the hook itself? Rather than theorize about them, I stuck some of them onto the outside of a window and made a video from the inside. This is what I learned.

The temptation (and you are sternly warned against doing this) is to pull the tab away from the surface, a rather obvious peel action that would seem to be no problem. I was puzzled why this required a force that is said to be large enough to remove paint from a wall. While setting up the video to see what really happens, it occurred to me that peeling cannot work because the hook itself is stopping the peeling action. You have, effectively, to peel the whole area of the rigid hook in one go, which is the equivalent of a butt pull, so it's no surprise that this will peel off the paint.

Instead, the act of pulling down on the tab causes all sorts of weird stresses within the tape itself. These would seem to be shear stresses, and we know that resistance to shear is much larger than resistance to peel. However, just as the lap shear joint taught us that a shear joint generally fails in peel, it is visually clear that the system starts to peel away gently at its edges. We are putting in a lot of force, much of which vainly tries to fail the system in shear. All it needs is for some fraction of that force to be distorted into peel and this rather conventional PSA should peel away with little effort.

The hooks come with a weight rating so I was interested to see what happened when the rating was exceeded. In my experiment the answer was "nothing" because even going to 3 kg, the 225 g hook remained firmly attached. Only when I applied a larger force did failure occur – in this case between the hook and the adhesive. Conveniently this left a nice strip stuck on the glass to test the idea in the previous paragraph. Indeed, with no hook in the way, a conventional peel removed the tape with no problem.

Why was the hook so massively under-specified? Why didn't they claim that it could stand 3 kg instead of 225 g? In addition to the natural fear of a lawsuit if some precious object weighing 226 g fell off, there is a necessary scientific caution. If the hook is always applied by a scientist onto clean glass at 20 °C and tested carefully, then 3 kg is a reasonable standard. Adding a latitude for poor application and the use of the hook in a cold winter or a hot summer, the claims have to be dialled back considerably.

6.10 ADHESIVES FOR PACKAGING

Spare a thought for those who have to make the adhesives for yogurt pot lids, packaged meats or fruits and just about any package that needs to be opened without attacking it with scissors or a knife. When the adhesives work well, which they do most of the time, we don't even think about them. When a package is rather too hard to peel open, we find it irritating, and if it is too easy to open, we worry that it might not have been sealed properly in the first place.

Because a sticky PSA (i.e. with a Dahlquist 0.3 MPa) would not be appropriate when these packages are opened, they are temperature sealed, i.e. hot melt PSAs which reach Dahlquist modulus at, say, 45 °C. Apart from that difference, the trade-offs are the same: they must work at a reasonable temperature range (fridge to room temperature) and, therefore, at a reasonable opening speed range (fast to slow). If, like me, you are curious about these things, try opening a part of a package while it is cold from the fridge, with a slow pull for one portion and a fast pull for a separate portion. Then wait till things are at room temperature and repeat the slow and fast tests. In an ideal world the adhesion would be the same under all those conditions. In the real world you might find a range of easy peels, stick-slip (raspy) peels, clean peels and peels with residue. You will mostly find that adhesive technology has become so sophisticated, thanks to the demands of brand owners and supermarkets, that you would have to go to greater temperature/time extremes to see unsatisfactory behaviour. This is all down to product evolution. At first, *any* peelable package was amazing and the first producers gained a large market share, until someone came up with a more controlled version, which brand managers selected in preference to the old type. Eventually, everyone caught up with the market leader and the cycle of improvement started again. What brand managers know is that one million people will *not* go onto social media to say that their package opened no problem, while one person with a problem pack will go viral and the million good packs are irrelevant.

Because the hot melt PSA trick works so well, and because a PSA can be tipped towards very strong adhesion when it doesn't have to accept the compromises of a typical tape, it is possible to make very strong bonds without the need for chemical crosslinking re-actions. This gives packaging manufacturers an interesting choice:

- Use a conventional polyurethane adhesive. Because it cures using ambient moisture, it can take 24 hours before the bond is strong enough to test. If you have spent 24 hours producing vast amounts of packaging and the test fails, it all has to be junked.
- Use a hot laminated PSA. In principle the bond is at full strength immediately (in practice, some time has to elapse because the high-speed process leaves the polymers rather too stressed) and can be tested before large amounts are manufactured. Because PSAs work across most materials, there is no need for extra chemistry to prime one or both of the surfaces. What's not to like?

Because of the perception that chemical bonds from "real chemistry" like polyurethanes must be better than the appar-ently weak bonds typical of PSA tapes, there has been a lot of resistance to the second approach. When I once suggested it to a company as a solution to a very difficult packaging problem, their reaction was that I must be mad. I think this is because PSA *tapes* must, by definition, be weak enough that they can be peeled and this gives the impression that all PSAs must be weak.

And once when a large corporation gave an excellent confer-ence talk about a new, green adhesive system for packaging, the phrase PSA was not mentioned once, even though it obviously worked via PSA-style dissipation. Over coffee I asked the speaker about this: "Shhhh! Of course it is a PSA, but if I call it that, no one will want it". Because adhesion is a property of the system, neither approach is perfect for all systems. By understanding the system and the ways in which PSAs react to temperature and time, the range of systems for which they are suitable in per-manent packaging is expanding.

Why would anyone want to stick two polymers together for packaging? First, we need to address why supermarkets have so much packaging. For fripperies like cosmetics and fancy choc-olates, the absurd level of packaging is caused by us, the con-sumers, erroneously associating a fancy package with a fancy product. Putting those to one side, we need to look at packaging from the supermarket's perspective. There are, as always, multiple factors:

1. Every μm of packaging material is a cost supermarkets would prefer not to have, in addition to be objectionable as waste plastic. There is intense pressure to "downsize" or not use packaging. Customers don't always appreciate this. A super-thin plastic lid may be a marvel of low-waste efficiency, but if it snaps (because it is thin) rather than peel back as expected, customers are irritated. If the adhesive is made lighter, a carelessly placed box of blueberries might burst open in the customer's bag, leading to even greater irritation.
2. There has to be a level of conservatism in the degree of protection because any customer complaint due to a faulty package is a large cost. In the UK a large fraction of supermarket waste used to come from leaking plastic milk bottles – the lids never closed reliably enough. To solve this problem, an extra plastic seal (applied via the induction sealing method described earlier) was introduced. It produces another tiny bit of waste and some inconvenience in having to pull it off, but is fully justified by the dramatic reduction in waste milk and waste bottles.
3. We consumers expect foods to be fresh and safe for far longer, so packs have to exclude oxygen and also keep all the food flavour molecules within the pack. The extraordinary longevity of modern packs of fresh fruit, for example, is thanks to controlled atmospheres (for example, high levels of nitrogen) inside the pack. For this to work, the pack has to keep the controlled atmosphere in as well as exclude bad things from outside.
4. We consumers expect convenience foods that can be, say, cooked within the pack in a microwave oven, or in a pan of boiling water.
5. The package has to be big enough to allow all the regulatory information to be printed in a size that is at least theoretically readable by consumers who want to know if the product contains, say, an allergen or is high in, say, fructose.

No single packaging polymer can deliver all those functions: PE easily lets citrus flavours out and oxygen in. PET is great for citrus flavours and is a reasonable gas barrier. LDPE is cheap and weak/floppy, PET is more expensive and strong/stiff. We also need extra barrier layers such as EVOH and aluminium. As a result, adhering layers together either during primary manufacture (co-extruding EVOH inside two layers of PE) or in a secondary process (adding a PE layer to a PET layer) is something done on a vast scale, with high efficiency (waste is expensive in many ways) and low cost.

The amazing thing about EVOH is that the thinnest of layers provides a good barrier. This means that most packaging films contain less than 5% of EVOH and the overall polymer is deemed by the regulatory authorities to be of recyclable grade. Purists would demand that the polymer should be 100% to make it fully recyclable, but this ignores the fact that all packaging contains residues from the previous contents such as foods, printed inks, paper label residues *etc.* so less than 5% of EVOH is not a significant burden.

Why do we carry on using non-biodegradable polymers such as PE and PET? The simple reason is that almost by definition, biodegradable polymers must be relatively weak and easily attacked by moisture – otherwise bacteria cannot get an efficient degradation process going. And the last things we want in most of our food packaging are weak polymers, easily attacked by moisture. Why don't we use bio-based polymers, from renewable sources? First, because those polymers so far available are not especially good polymers for packaging – you would need more of them to provide the same protection, which is not a sustainable approach. Second, because it is not obvious that diverting arable land from food crops to growing chemicals for plastics is really great for the planet. Early, overenthusiastic analyses of the trade-offs between petro- and bio-based materials under-estimated this "land use change" factor (its formal name in Life Cycle Analyses). More recent EU standards for comparing petro- and bio-based materials have much more realistic values which often, to the surprise of many, means that petro-based materials are *better* for the environment than their bio-based equivalents. Third, because the bio-based polymers currently available are not "biodegradable" and are merely "compostable" but not to a level that makes them good for municipal composting plants. This fact means that in many countries, these polymers are *not allowed* in the composting waste stream.

This diversion away from the science of adhesion is deliberate. It is very easy to be dismissive about packaging and critical of the complexity of many packaging materials. The naïve call for biodegradable and single-polymer packaging material does not take into account the prime function of packaging which is to protect. The adhesion part of this overall protection is generally unknown and, therefore, unappreciated. The key to saving the planet is to focus precious resources on those activities that will make a substantial difference. This can only happen when there is informed debate. The people I know in the packaging industry are not dismayed by calls to reduce packaging waste, indeed, because they live on the same planet as the rest of us, they are keen to reduce packaging waste. What dismays them is *ignorant* calls that divert resources from actions that will make a difference to actions that waste even more resource.

CHAPTER 7

Sticking Other Things Together

Rather than risk being kicked out of a large hardware store, I tracked down the store manager to ask her permission to spend an hour or so in the adhesives section with my smartphone for looking up safety datasheets and a notepad to write things down. She was happy to accept my explanation and two hours later I emerged with lots of new information and ideas for this book – and lots of samples of glues and tapes for testing further at home. The reason for my visit was a realization that, although I say that "adhesion is a property of the system", I was very far from knowing all the systems that need to be stuck together. It is true that if someone wanted to be able to stick most things together adequately, a small selection of adhesives would be all they required. That's what the previous chapters have been about. A tube each of solvent-based glue, superglue, epoxy, urethane and PVA wood glue would cover most tasks adequately – indeed, that is what I have in my toolbox. However, when I asked expert users (acknowledged in the Preface) about the adhesives they used regularly, I was told that there were jobs which required specific adhesives, and that is what this chapter is about, and is why I have lots of unusual adhesives stored away in a cupboard.

The hours in the store indeed revealed a range of tasks for which the general-purpose solutions would not be suitable and where there were adhesives designed for the specific job. I am stretching the word "adhesive" by including cement and ceramics. This is

Sticking Together: The Science of Adhesion
By Steven Abbott
© Steven Abbott 2020
Published by the Royal Society of Chemistry, www.rsc.org

because the principles are no different and the science fits in with the "strong adhesion" theme that silanes (-TMS and -TES) are rather good general-purpose adhesion promoters.

The legitimately large range of products is confused by the pseudo-ranges created by marketing. A lot of my time in the store was taken up by trying to distinguish the genuine solutions to real problems from marketing solutions to problems that don't exist.

An interesting observation is that many adhesives manufacturers contradict their own claims. If their Wonder-Adhesive which "sticks anything to anything" was as good as they imply, why do they also sell other types of adhesive? The scientific answer is that there is no such thing as a Wonder-Adhesive. The marketing stance is, I suppose, to hope that no one notices their self-contradiction.

While no single reader will be greatly interested in all the adhesion challenges described here, I am confident that each challenge will have at least one reader to whom it is important. If your specific adhesive challenge is not included, that may be because I did not spend enough time in the store, or because it has already been covered earlier, for example in discussions of the science of superglue or epoxies, or of the various possibilities of PSAs.

Most of the time we want strong adhesion. The challenge of creating controlled low adhesion is addressed at the end of the chapter.

We start with the apparently humble, but actually super high-tech challenge of sticking bits of sand and stones together.

7.1 CEMENT AND CONCRETE

I once had to research a high-tech ceramics application and was very surprised to find that many of the best scientific papers of relevance came from workers associated with the cement industry. Like most people I had taken this humble material for granted. It's just some grey stuff you mix with sand, gravel and water to make a solid lump (concrete) which you then forget about. I was entirely wrong in this view. The cement industry is highly sophisticated because it has to be. As mankind's single most used material, even a 1% advantage in cost or performance or CO_2 reduction is of huge significance. Purists will want to distinguish between cement, mortar and concrete. Because I am focussing on the adhesion part which is driven by the cement chemistry, I will use "cement" as a generic term.

The simplest, and least interesting, cement is called "non-hydraulic" because it only sets when the water in which it is supplied dries out. Limestone, calcium carbonate, is heated to produce calcium oxide, emitting a molecule of carbon dioxide. Mixed with water the calcium oxide produces "slaked lime", calcium hydroxide. If this wet cement is mixed with sand and gravel, as the water disappears during the setting process, the calcium hydroxide reacts with CO_2 in the air, returning to the original calcium carbonate, creating a reasonably strong matrix.

This sort of cement has no good reason to be especially well-adhered to the sand and gravel included to create concrete. If mixed well, without too much air getting in the way, then there is good surface contact, the minimum required for strength. The lime is somewhat aggressive so there might be some bonding reactions with the surface of the particles to create a little extra strength, Overall, however, the inevitable presence of voids as the water dries out means that there are crack initiation sites within the structure (remember Griffith's law from Chapter 2). The result is that the strength of non-hydraulic cement is merely adequate.

The real cement industry is based around hydraulic cement that requires water in order to set, and can therefore be used under water if necessary. It is common to point out how smart the Romans were to invent such a cement. It wasn't so much smart as lucky. It had been known for centuries before the Romans that you could make cement by "burning" (firing) lime together with things we now know to be silicates such as clays. The Romans were lucky because the silicates available in large quantities from past eruptions of Mount Vesuvius had an excellent balance of setting speed (even under water) and ultimate strength and resistance.

We found in Chapter 5 that the reason silicates are so important for a good adhesion is that they provided anchor points for reactive "tails" in adhesives to latch onto. In many conventional adhesives, the adhesion promoters (-TES and -TMS) are silicate based. Silicates are able to react with many surfaces containing oxide and hydroxide groups, which, in a world where the atmosphere is 20% oxygen, means most surfaces. The sands and gravels added to create concrete contain silicates plus (remembering the example of the old-fashioned bath in Chapter 5) oxides of aluminium and iron which also react with silicates.

Isn't sand just sand? Unfortunately not, and there are even from time to time reports of sand shortages and sand mafias stealing whole beaches of the right sort of sand. Sands from coral islands, for example, are unusable for good concrete as they are mostly calcium carbonate. It turns out that the right shape/size/ silicate content of a good sand for cement is surprisingly hard to find in the huge quantities required.

A good cement contains a finely ground mix of calcium oxide and a reactive form of silica, plus, in many cases, reactive alumina. Mixing ratios can be selected in trade-offs of strength and setting speed. Water reacts to create calcium silicates (and aluminates) which in turn can react (because of the high alkaline pH) with the surfaces of the sand and gravel to form further complex silicates/aluminates in a continuous network (Figure 7.1).

It took a large part of the 19th century to work out the basic science of creating what is now a standard "Portland cement" with a good balance of cure speed and strength/endurance in use, i.e. to produce via science what the Romans achieved through good luck. Although the basics have not changed all that much since then, there are many specialist requirements that continue to push cement technology.

An important issue for all powders/particles dispersed in a liquid such as water is how to have the easiest flow with the least amount of liquid. As the fraction of solids increases, the viscosity

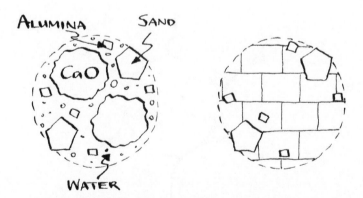

Figure 7.1   A loose mix of cement particles (calcium oxide), alumina and sand when reacted with water creates a solid network of calcium silicates, aluminates etc. where everything is locked together.

starts to get very high, making the dispersion unusable. For cement this is important because any excess water (beyond that required to complete the reaction) has to disappear over time, slowing down the building process; it also means lifting a lot of unnecessary water to the top of a skyscraper under construction. There are two tricks:

1. Make sure that the cement is made of three different sizes of particles (Figure 7.2). If you have just one size of large particles, when they pack together viscosity rises rapidly as the particles approach 63% of the volume. However, there is still plenty of space between the larger particles into which smaller (then even smaller) can fit, taking the maximum packing to 72%. That is a very useful increase.

2. Add a plasticizer. This is a molecule that sits at the surface of each particle and reduces the particle-to-particle attraction that would otherwise cause a high viscosity. There is a low-tech version which is a few drops from a bottle of dishwashing liquid that many builders keep in their vans for cement emergencies. The internet tells us that dishwashing liquid is bad and a proper plasticizer should be used. I tracked down what this is. It is sodium laureth sulfate, a well-known surfactant commonly used in dishwashing liquids.

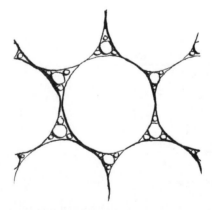

Figure 7.2   To pack as much cement as possible into the mix, smaller particles sit in the spaces between larger particles.

3. Add a superdispersant. These are more sophisticated versions of the plasticizers, with careful optimization for large-scale construction projects.

7.2 CERAMICS

Ceramics, from humble pots and dishes through to sophisticated turbine blades on jet engines, are stuck together via a rather surprising process.

The starting point is a room-temperature paste that is given the desired shape. Clays are made plastic by their water content and need nothing extra to give them the required "green strength", i.e. the strength to keep their original shape when handled carefully. Technical ceramics typically contain polymers and solvents during the forming process. When the solvents evaporate, the polymer acts as a contact adhesive keeping the ceramic particles in place.

In both cases, the need is for the highest possible fraction of ceramic particles while maintaining processability. This is the same challenge as for cements and the same tricks are used: multi-sized particles and special dispersants that reduce the particle-to-particle interactions. Any excess material is a problem during the drying/setting process and during the final firing process, so it is important to pack as much ceramic powder as possible into the green form.

For ceramics held together by polymers, the first stage in firing is to burn off the polymer (and any dispersants that might still be in there, since we now want to *encourage* particle interactions) at a few hundred degrees. This is a dangerous time for the ceramic because it is losing its structural support before it has a chance to form its own particle-to-particle bonds.

What turns the green form into a tough ceramic at higher temperatures is the curious phenomenon of sintering (Figure 7.3). This is discussed in more detail in the 3D printing Chapter 9. What happens is that although the particles do not melt, the atoms at the surface are so unhappy about being exposed (they have a high surface energy because they would rather be in the bulk of the ceramic) that they will spontaneously hop along the surfaces and create tiny bridges between particles; these bridges can slowly grow to form strong connections.

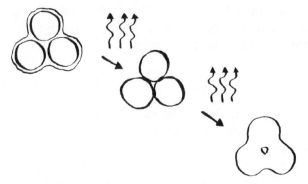

Figure 7.3 The ceramic particles are held together with a polymer which disappears when heated, producing a delicate assembly of particles that sinter together at higher temperatures.

A ceramic material which might require 3000 °C to melt will sinter quite happily at 1500 °C. Often the process needs some encouragement via various added "fluxes" which have relatively low melting points and can therefore flow (as their name suggests) and speed up the sintering by part-dissolving surface molecules which can then move faster and further.

The clay pots, stoneware, porcelain and bone china we know are all based on silicates – as with Portland and Roman cements. It is the same ability of silicates to form linked networks that creates the final ceramic. The differences lie in the firing temperatures, fluxes, the precise ingredients, and their purity (you need fine kaolin for porcelain and ordinary clay for a pot). For example, roasted bones add phosphates to kaolin-type clays, allowing the creation of bone china. The (Western) inventors of the process were trying to create china of the super-high quality found in China but had no access to the specific (and rare) form of clay used for the Emperors. The phosphate from the bones happened to give the same toughness and clarity as the original clay from China.

There is one more vital difference between different ceramics which we can understand from our adhesion science theory. Any holes/pores in the fired clay are sites for cracks to start. Porcelain is tougher than stoneware partly because of the more sophisticated chemistry and higher firing temperatures, and partly because the clay particles themselves are much finer. Finer particles, semi-fused together, leave fewer large holes between them, and therefore fewer Griffith's law crack sites. The fine particles also scatter light far less, so porcelain is translucent. Because the porcelain is much tougher, it can be made thinner, giving a further boost to its translucence.

7.3 HOLDING ON TO BRICKS AND CONCRETE

There are times when something, for example some hoist gear or a ceiling panel for a tunnel, has to be attached strongly to a brick or concrete structure. A strong bolt is often necessary, going into a drilled hole. Then the question is how the bolt stays attached to the structural material which can be rather brittle. A strong adhesive is needed that wraps itself around the bolt, fills the gap and is in perfect contact with the material itself. A strong epoxy is one of the standard ways to do this, using an adhesive gun to mix the two parts of the epoxy and squirt it into the hole before putting in the bolt. An alternative is to use a double-compartment glass cartridge which fits nicely into the hole and contains the two parts of the epoxy. The bolt is "screwed" into the hole (shattering the glass if that was used) to ensure that the epoxy flows into the threads to provide some (in this case genuine) mechanical interlocking once the epoxy is cured. These epoxies are usually filled with particles so they will shrink even less than a standard epoxy, helping to ensure that there are fewer built-in gaps and stresses. The epoxy itself can react slightly with the silicate surfaces of the brick and concrete, providing entanglement across the interface for further resistance. Done right, this is a reliably strong system. As discussed in Chapter 5, the Boston Big Dig tunnel ceiling panel failure was due to an epoxy that had not cured sufficiently, allowing the bolt to creep over time.

An alternative is a "vinyl polyester" which is a broad term that can include an epoxy component as well as a polyester bulk polymer with a good balance of strength and flexibility. The "vinyl" part contains reactive double bonds that cure via a radical mechanism, itself initiated by something like a peroxide. The need for the catalyst means that the system is also two-part.

Another variant is to use acrylates, which we met in terms of UV curing. As you cannot get light into these systems, they too are two-part systems requiring a peroxide initiator because acrylates contain the same reactive double bonds as vinyls.

In many of these heavy-duty systems, the filler is Portland cement, which means that any moisture that gets into the fitting (or is provided in the two-part delivery system) helps create silicate bonds to the brick or concrete (Figure 7.4).

The trick in designing all these systems is, as discussed previously, that whether they are using epoxies, vinyls, polyesters or acrylates, there is a balance between 2-functional, 3-functional and 4-functional groups. A system with only 2-functional groups will be flexible and weak. A system with only 4-functional groups will be rigid and strong, though not tough. The formulators have to strike a balance of functionality, often mixing 2-, 3- and 4-functional materials and then doing extensive testing for different sorts of potential problems. An adhesive which is wonderful for very heavy, constant loads in a warm environment might perform poorly for a bolt that is being loaded and unloaded (e.g. as part of a hoist) in a cold, damp environment.

The experience of every adhesive formulator is that getting a "good" adhesive is straightforward. Much harder is making sure that it works in a wide variety of environments when applied by people with very different levels of skill. This requires a lot of testing work. If the bolt is holding up a key part of a tunnel or is the main support for a hoist, failure is more serious than a failure of a household adhesive. The materials in a tube of these adhesives are not all that expensive. The cost of proving that the adhesive (and each batch of it) is stronger than the brick or cement surrounding it is a large part of the price you pay for a quality product.

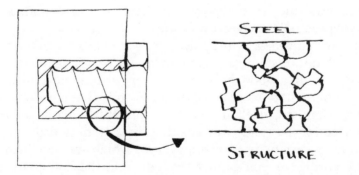

Figure 7.4   For fixing heavy items to structures, a bolt is screwed into the adhesive in a hole. The adhesive often contains Portland cement particles that can lock on to the cement or brick surface.

7.4 PLASTERBOARD/DRYWALL

Plasterboard or drywall, depending on where you live, is a partially-hydrated calcium sulfate ("hemi-hydrate") which becomes further hydrated upon contact with water, forming a crystalline network of gypsum ("dihydrate"). The solid gypsum is far too heavy and far too brittle to be of use, so it is mixed with paper or other forms of fibre and highly aerated to create a board which is full of holes when dry, making it light enough for use. Because of these holes and Griffith's law, plasterboard is easily cracked. But since users are aware of this, and because plasterboard is not used to provide any serious mechanical strength, the balance of holes (lightness) and solidity (strength) is acceptable.

The adhesive is basically the same material (usually with some plasticizer to make it less brittle) because during the adhesive curing process the plasterboard itself becomes wet and soggy, allowing the network to form perfectly as the water converts the hemi-hydrated sulfate into the di-hydrated form.

Creating an even coating of the adhesive across an entire sheet of plasterboard is difficult, unnecessary and probably less satisfactory overall because it is hard to correct for unevenness of adhesive thickness. The classic technique is "dot and dab" with blobs of adhesive thrown from a trowel onto the wall and then squashed by the board. The blobs will be uneven, but because there is space between the blobs it is easy to get them to flow out under the pressure needed to set the board square and vertical (Figure 7.5). This is Stefan's law, discussed in the Chapter 2, in action.

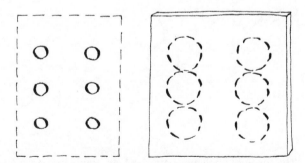

Figure 7.5 Dots of plasterboard cement are placed onto the wall then the board is placed on top. The dots expand, but because they are free, it's possible to apply more or less pressure to ensure that the plaster is aligned properly with the wall.

7.5 TILES

Tiles are ceramics and if they are being stuck to brick or concrete, which are also ceramics, it makes sense for the tile adhesive ("mortar") to be like cement itself, relying on silicate reactions to form adequate bonding across the interface. Sticking to plasterboard and to wood is not a serious challenge as the liquid adhesive can flow into and around the fibres.

We can imagine a dry tile adhesive powder being a simple cement which is mixed with water and sets as normal. However, pure cement adhesives lack some key features for general use, so modern packs of tile adhesives contain multiple ingredients. Especially important are polymers that provide the necessary flexibility to accommodate the normal movements of building materials as they respond to the weather, minor settling, foot traffic (tiles on a suspended floor) or anything else that creates a mismatch with the rigid tile.

A typical example of a polymer in tile adhesives is EVA (polyethylenevinylacetate), which is similar to the PVA in wood glue but with some of the desirable properties of polyethylene. In wood glue, the water-insoluble PVA emulsion particles are kept apart by stabilizing molecules covering the surface. As the water dries, the particles come together (coalesce) into a continuous, tough film (Figure 7.6). In tile adhesives, the EVA particles are

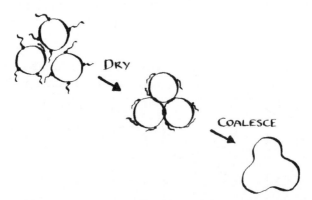

Figure 7.6   The insoluble polymer particles have stabilizing polymers to keep them apart in the aqueous environment. As the water disappears, the particles clump together then coalesce into a single, insoluble, particle.

produced in water and then spray dried into a powder that is mixed in with the cement. As water is added to cure the cement, the polymer particles are nicely dispersed within the wet adhesive, then coalesce to form a tough polymer network adding the required flexibility to the adhesive.

The polymer particles perform an additional function – they fill in the cracks between the cement particles, giving a much tougher film. In the days before these "redispersible polymers" it was necessary to apply a thick layer of cement to get adequate strength across the bond. With polymers the bonds need only be 3 mm thick.

There are many more ingredients in a tile adhesive. Amateurs (or those working in hot climates) might need retarders, things like sodium citrate, to slow down the cement reaction. Professionals can use tile adhesives that are hardened in 20 minutes thanks to the use of accelerators such as calcium formate. Such fast-setting adhesives mean that they can quickly apply the grout (filling the spaces between tiles), allowing that part of the job to be fully finished so that the next task in the build project can be worked on. The danger with fast setting is that the adhesive may not have had time to fully create a solid network. Water-retaining additives are required to help complete the full network-building reaction over the ensuing hours or days. A typical additive is MHEC (methyl hydroxyethyl cellulose) which is hygroscopic (attracts water) and also creates a rather sticky/viscous solution with the water in the gaps so that it doesn't flow out too rapidly as other parts dry out.

For tiles, there is the same problem as for plasterboard: applying an even, thin layer is hard, making it difficult to adjust the tiles. The trick with tiles is to apply the adhesive with a comb (Figure 7.7). This makes it much easier for thick portions to flow out fully (no comb marks), with the thin portions having comb marks but still with plenty of contact for strong adhesion. The best way to achieve this is via the "slide" technique that adds shear to smooth out the comb lines. The shear can temporarily reduce the viscosity, allowing the adhesive to flow. Air can escape along the grooves created by the comb.

When tiling onto plasterboard there is the danger of too much water being absorbed into the board, which reduces its availability for the reaction. Conversely, adhesion can also end up

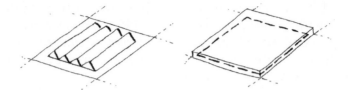

Figure 7.7   By combing the tile cement, it makes it easy to spread out the cement evenly across the whole tile, with an added twist to help the cement to flow. Without the comb, it would be impossible to get a level tile.

being too good, in the case of a tile having to be removed for any reason. A pre-coat of water-impermeable PVA glue solves both problems.

Premixed liquid tile cements obviously cannot use the classic cement chemistry created when water is added. Instead they make use of co-polymer adhesives (delivered in an emulsion), which, through careful choosing of the mixing ratio, allow the formulator to tune the properties to match the application. Co-polymers are typically some variant of a styrene-acrylate mix, with fillers such as calcium carbonate which provide a cheap source of strength and reduction in the amount of water needed. A pure polystyrene (too fragile) or polyacrylate (too soft) would be useless as a cement, while the right ratio of the two gives the required balance of strength, flexibility and toughness. The problem with these pre-mixed adhesives is that the water has to be able to escape. This isn't so difficult to manage on plaster or wood, but can be a real challenge on more impervious structures for which drying times might be hours or days.

We can now move to a different class of adhesives based on solvents rather than water.

7.6   RUBBERS AND BITUMENS

These are discussed together merely for convenience because they are each nicely soluble in hydrocarbon solvents such as the large range of naphthas, which are distilled (and often purified) fractions from oil. Naphthas are available in a wide range of volatilities, from high to low, that can be selected according to the specific application. Rubbers and bitumens themselves are also rather sticky materials that become stickier in the presence of the naptha solvents.

To stick bituminous felt tiles requires (often) a rubber adhesive, but you could equally stick bits of felt together with just the naptha. It partially dissolves the bitumen, allowing the molecules to intermingle and entangle as the solvent evaporates. Although the solvent is trapped between the layers, it can readily diffuse through the bitumen allowing the rubber-bitumen or the direct bitumen bond to reach its normal no-solvent strength.

Rubber adhesives are common for things like carpets because they are free-flowing enough when sprayed from a can to wrap around the fibres, rapidly become sticky as the solvent evaporates for a quick tack, then become strong, sticky solids once the solvent has completely disappeared – which it can manage easily by travelling through the carpet. They also work for vinyl flooring and tiles because the solvent can migrate fast enough through the vinyl to allow good drying over many hours. The rubber is also desirable because it can accommodate the flexing as people and objects move over the surface.

The general-purpose home adhesives based on rubber work in the same manner – the solvent in a layer applied to both adherends is allowed to mostly evaporate so that the parts can be joined to give an adequate bond. As the solvent diffuses out of the bond, the adhesion gets stronger. This works well on materials like leather and with more difficulty on non-porous materials.

Similar styles of adhesives based on different polymers use different volatile solvents such as methyl ethyl ketone, acetone or ethyl acetate, which feature in our next topic. Although the specifics of each case are different, the general principles are the same.

These flexible, solvent-based adhesives do a good job but have the obvious downside of the solvent fumes. For professionals who know what they are doing, and for the home user doing the occasional job, the risks are small. The quantity of Volatile Organic Compounds being released is insignificant on a global scale. Still, with there being many alternative adhesives for many of their jobs, this class will gradually fade away.

7.7 NAIL POLISH

One characteristic of my generation is that men don't usually apply nail polish. So, when I phoned a professional nail expert asking for a home visit, I felt compelled to assure her that I was just a mad scientist and that my wife would be around when she came. It was a fascinating experience and I was deeply impressed with her skill, the technology and the sophistication of the formulations being used. I was also very glad that the session was planned so that my nails could be restored to (my) normality by the end.

The reason for the session was that I had been asked to help design a new "eco-friendly" nail polish and I had to understand the current technology before thinking up ideas for a new one.

The basics are common themes:

- Prepare a clean surface on the nails by removing any junk (light buffing) and using a solvent to remove any oils.
- Then it's good to apply an adhesion promoter. The surface of the nail contains a variety of acidic and basic sites, making it relatively easy to provide a molecule that at one end binds to the nail and at the other is compatible with or reactive with the polish itself.
- Then apply the polish as a thin coat, wait till it is dried/cured and apply a second thin coat.
- Add a final gloss protective coat. Any slight damage to this will not show up as much as damage to a coloured layer.

For the main coloured layer, why not apply a single thick coat? The enemy of all adhesion is internal stresses built into the coating, because any external stresses (that is, in most cases, normal usage) are additional to this higher stress load and can perhaps trigger a crack. A thick layer of a solvent-based polish will dry at the surface. It can readily shrink as it floats on the liquid beneath it. As the solvent within the coating subsequently evaporates, the top surface finds itself on a shrinking background. Either it is soft enough to bend into wrinkles, or it remains smooth, with built-in stresses.

There is a different problem with the UV systems. The top cures first because it sees more UV light and might build up stresses. In addition, with a thick coating and a highly-coloured polish, the UV light might not penetrate far enough to give a good cure at the nail surface.

The solvents used are the ketones (acetone and its less volatile equivalent methyl ethyl ketone) and the acetates (ethyl acetate and its less volatile equivalent, butyl acetate). The acetates are often seen as "friendlier" than the ketones. Other, greener solvents are available (such as ethyl lactate) and are fully effective. They are less popular because they are less volatile, creating the problem that the time taken to produce perfect nails is too long for most customers. Although formulators can readily create adequate formulations with water as a solvent, water is surprisingly hard to evaporate and so is similarly less popular with impatient users and time-pressed nail salons.

The simple polishes are usually based on nitrocellulose which has a good balance of solubility in the solvents, ability to disperse lots of pigments and to produce a hard, smooth surface.

The family of "acrylic" or "gel" polishes uses the same acrylate chemistry as in UV polishes, but with something like benzoyl peroxide as the radical-producing initiator, as described earlier in the case of acrylate bolt-bonding adhesives. Then we have the UV polishes where the UV light hits a photoinitiator which triggers the reaction by producing radicals. In both cases, an adhesion promoter will contain an acid or base to bind to the nail along with an acrylate that reacts into the overall system giving us entanglement at the nail interface.

Whatever the system, there is the usual trade-off of toughness versus hardness. Nitrocellulose on its own is rather too hard so various plasticizers are added to soften it a little. For the acrylate-based system the number and type of 2-, 3- and 4-substituted acrylates are varied to get the right properties.

All good things come to an end, and nail polish has to be removed at some time. Having my finger soaked with an acetone-filled pad wrapped in aluminium foil to stop evaporation was a strange experience. If I ever did that in a chemistry lab, I would be reprimanded for contravening safety practices. If there is risk of contact with any solvent, good lab practice is to wear appropriate gloves.

And I would certainly be thrown out of most labs if I put acrylate systems, benzoyl peroxide or UV initiators anywhere on my unprotected hands. Somehow the needs of fashion outweigh standard good laboratory practices. Fortunately, those who work all day with these chemical systems in the better nail salons are trained on the risks and safety precautions, and the risks from occasional exposure to home nail care products are not excessive.

The chemicals used in these processes are also efficient at extracting natural flexibilizing molecules from the nails, which can become brittle over time. The final stage of my treatment was the application of some oil to soak into the nails and, hopefully, repair some of the damage from the chemicals.

7.8   LIPSTICK

Because I consult for the cosmetics industry on the science of skin creams, I was once asked if I would like to help solve some problems about lipstick. My daughter and daughter-in-law heroically volunteered to teach me about the topic. Applying lipstick wasn't too bad, but trying out lip gloss in the cause of science was a harrowing experience. I was relieved when I could finally wash it all off. After that bit of practical science, and after reading the scientific literature, I reluctantly concluded that I had nothing novel to contribute to lipstick science. The basic formula of wax, pigments and oils (to tune the rigidity of the wax) seems to be hard to beat.

The ideal product is one that just works, with no extra effort or steps. The brightly coloured material must be applied easily, stick perfectly, never transfer to another object on contact and remain in place during a day. The science of adhesion tells us that if that was the challenge then it would be tough but do-able. The final requirement is that it should be removable on demand and leave no trace of the original colour on the lips. Without some pre- and post-steps, that combination of properties seems to me to be impossible.

If we allow some preparation steps, our familiar adhesion science tells us what to do:

- Make the lips as smooth as possible via exfoliation of dry skin and moisturization – rough lips will be sites of crack propagation.

- Remove oils from the surface – these will be weak spots.
- Remove water from the surface – the waxy lipstick does not like it.
- Apply a primer – we can readily imagine a molecule with one end that loves to attach to groups on the surface of the skin and a long waxy tail to intermingle with the lipstick. Because the primer is colourless and we have only intermingling with the lipstick itself, the colour can later be removed without staining the lips.
- Apply the thinnest possible layer – a thick layer is more easily pulled off. To obtain the correct colour intensity, the pigments must be ground to an optimal small particle size.
- Because it is hard to apply a perfect thin layer, apply a second thin layer on top of the first. It seems to be common practice to dab the first layer with a tissue, though I cannot identify the science behind this. Maybe it is a bit of Stefan's squeeze to push lipstick into remaining holes or, the opposite, removal of unhelpful excess.
- Apply a protective transparent overcoat, which might be some sort of solvent-based system that dries sufficiently hard to be protective and flexible to be tough.

7.9 METAL TO METAL

We have already discussed how to use liquid metal (solder) to join some metals (especially copper) together. Welding has also been briefly mentioned. Although there are sophisticated techniques for specialist metals, welding is most straightforward with steel where the steel to be joined is melted along with extra steel (e.g. as the welding rod) which contains (as with solder) a flux to remove oxygen from the surface. As the steel solidifies it basically becomes a single piece of metal and holds no further adhesion puzzles. Brazing is an intermediate technique that joins high-melting metals like steel with medium-melting metals like copper. Gold and silver are readily adhered via similar processes such as soldering.

Most adhesion of metals involves the sorts of serious structures where metals provide strength, so if welding is not appropriate, the adhesive should also be strong. The default, therefore, is a strong epoxy. At one time, I had to read a lot of the literature on how to create a strong bond at the metal interface. There was a surprising amount of bad science and misunderstanding. In reality it is all very simple. As much junk as possible needs to be removed from the metal surface, allowing it to be a simple surface with a thin layer of oxide which is inevitable for most metals. An exception is iron/steel where a phosphate surface is sometimes desired for long-term corrosion resistance.

With such a clean, controlled surface, the adhesive either reacts directly with the –OH groups on the surface or contains an oxide-loving group such as the much-discussed silanes, -TMS and -TES at one end of an adhesion promoter – with the other end ready to react into the adhesive. An amine group at the other end is ideal for reacting into epoxies. For iron/steel, a phosphate group is especially effective instead of a silane, providing adhesion promotion as well as the corrosion resistance.

Because a highly crosslinked epoxy can be dangerously brittle, the core–shell rubber trick, discussed in Chapter 5, is used to add a bit of crack-absorbing rubber surrounded by a shell that reacts into the epoxy system.

When the bond doesn't require huge strength, the SMP types (silane modified polymers, silane hybrids) discussed in Chapter 5 have a good chance of working well because the silanes used to polymerize will also provide adequate bonding to most reasonably clean metal oxide surfaces.

After that, any household glue, especially the cyanoacrylates, will do the job if the surfaces are smooth and flat enough – allowing the thinnest possible glue layer to be used. A thick layer is more exposed to water and chemicals and, as we have seen before, in terms of resistance to butt pulls, thinner is better. From Stefan's squeeze we know to add ten small drops of adhesive rather than the equivalent amount in one big drop. Cyanoacrylates' resistance to shear and peel is so limited that they really should not be used if those failure modes are a serious risk for the joint.

7.10   WOOD TO WOOD

We have already discussed some of the wood-to-wood adhesion issues. Animal glues are excellent for jobs where removability is an advantage; they are also used in special techniques such as veneering where heat can be applied to melt the glue if necessary.

PVA glues have the distinct advantage of being relatively flexible so they can cope with the expansion and contraction of joints as the moisture content of the wood changes during the year. The water in the glue disappears by a combination of evaporation and absorption into the wood, with some tendency for the small PVA polymer spheres to be drawn into and around the fibres. As the water disappears, the polymer spheres coalesce into a water-resistant polymer. That's the official story.

The real story is that polyvinylacetate (PVA), which is genuinely water insoluble, always has a small percentage of water-soluble polyvinylalcohol (PVOH) within the polymer and (usually) as an additive to help it remain dispersed in the water, plus some surfactants. Right from the start, these modified PVAs come with a susceptibility to reduction in bonding strength when exposed to water vapour (humidity) or water itself. Although, the hydroxyl groups in the PVOH naturally provide reasonable anchor points to the wood, increasing adhesion, they provide weak points when water is present.

PVA is, therefore, only somewhat water resistant. It can be made more water resistant through co-polymerization with other monomers, creating a more hydrophobic polymer and/or by adding chemicals that, while the adhesive is drying out, react with the hydroxyl groups in the PVOH to render them less susceptible to water. The danger with both approaches is that the many good points of the original adhesive are compromised. For example, some reactions to reduce the –OH content tend to increase the viscosity of the adhesive which, in turn, makes it harder to squeeze out excess as the joint is formed and also will tend to reduce the ability for the adhesive to flow around fibres. I get the impression that for those who really need water resistance, it is better to use a different adhesive in the first place. If the wood is going to be wet most of the time then PVA's selling point, its ability to resist cycles of humidity, is no longer relevant.

Water-based glues are a problem for joints that involve end grains; the wood can absorb a lot of moisture along this exposed area so a tight joint will shrink a lot as the water disappears. For these sorts of joints, the polyurethanes or superglues seem to be preferred.

For real water resistance, i.e. for marine applications, the urea/formaldehydes already discussed are fine, as are the variants with melamine or with resorcinol.

We have already discussed some of the plusses and minuses of polyurethane glues. The fact that they are water activated means that the water in the wood provides good activation at the interfaces, with the downside that excess water creates foams which can (helpfully) fill voids and (unhelpfully) create relatively weak spots. Polyurethane chemistry is so versatile that glues of different setting times can be created, allowing the skilled woodworker to get the right balance between speed for routine, high volume jobs, and repositioning time for more complex jobs. The simplest way to be able to sell a 5 min version and a 30 min version of a polyurethane with otherwise identical properties is to put 6× less catalyst into the slower version.

Epoxies have often been investigated in the scientific literature as they should in many ways be ideal for woodworking – for example the –OH groups in the wood should nicely participate in the epoxy reaction. Yet they are generally shunned, partly because their (generally) high viscosities make them tricky to provide desirably thin layers but mostly because their rigidity makes them unhappy partners with the ever-changing wood.

The superglues are great for skilled woodworkers and not so good for those amateurs who cannot create joints and surfaces with high accuracies, and who can't get things aligned right first time. Superglues are so poor in terms of mechanical and water-resistance properties that any significant thickness is a weak spot. Because they set rapidly, aided by the moisture in the wood, any hesitation means a botched job. For precision joints and repairs put together by experts in seconds with the thinnest possible layer of glue, they are great – as long as the adhesion task does not ask for any peel or shear resistance for which the superglues, as mentioned several times, are known to be unsuitable. A typical example is the fancy mitres that go around cabinets. They are purely decorative, needing no great strength, while requiring the thinnest possible layer of adhesive for the finest possible join. With an accelerator spray on one side and the cyanoacrylate on the other, the joint is ready in a few seconds. The heat generated by the accelerated reaction is readily detectable as the joint is made.

For all glues with setting/curing times it is normal to clamp the joint. It is worth remembering Stefan's squeeze law which tells us that halving the original thickness of adhesive might be fast, but halving it again will take $8\times$ longer. As the clamp is applied it is easy to imagine that it has done its work when a satisfying amount of adhesive appears, ready to be wiped away. If this took 30 s, it is hard to imagine that it will be another 4 min before the next 25% of the adhesive comes out. The situation is worse than that. The initial flow of adhesive will reduce the pressure, so without a re-tightening (or, equivalently, a constant pressure clamping device) the excess will take even longer to flow out.

For those readers who want to make skyscrapers out of wood ("plyscrapers"), the starting point is CLT, Cross-Laminated Timber. These are multi-layer laminates of pre-dried, precision-cut wood from which all major imperfections have been removed, with each layer at $90°$ to the previous one. The layers are coated with adhesive then bonded/cured under high pressure. The adhesive everyone uses is a polyurethane that cures via initiation from the water in the wood, with the ability of a small percentage of wood –OH groups linking to the urethane to give tough entanglement. There are no problems of formaldehyde emissions typical of older styles of synthetic woods which often used systems such as melamine/urea/formaldehyde. Because CLT is still, basically, wood, it is only usable as the core of a building (replacing steel and/or concrete), above the damp-proof level and protected by external cladding. The polyurethane is, therefore, not exposed to especially harsh conditions and the lifetime of the adhesive matches that of the wood itself. To answer the big question, yes, they have done all the fire tests – CLT is fine because the outside chars, insulating the inside and providing the required time to failure.

The common assumption throughout this book is that thin layers of adhesives are provided in exactly the right places. Woodworkers who could make precision joints had no problem with this, though such joints are, necessarily, slower to create, and difficult to make in materials such as chipboard. An alternative approach is to make a somewhat sloppy joint with plenty of adhesive and find some mechanism to tighten up the joint. The biscuit joint does this admirably. A special saw cuts an elliptical groove into each of the two pieces of wood to be joined and adhesive is placed in both. A "biscuit" of highly-dried/compressed wood (typically beech) is slotted into one groove containing the right amount of adhesive and its other half is mated with the second, similar groove. The trick is that the biscuit absorbs the water from the glue (typically PVA) and expands to create the desirable thin layer between the surfaces. From my own measurements of dunking some biscuits, the expansion seems to be modest, just a few %. That is enough to make a difference. Adhesion is a property of the system, and in this system there is an ingenious dynamic element creating the overall strong adhesion from an otherwise imprecise setup. For the trick to work properly, the biscuit must be dry and un-swollen at the start. Those who leave their biscuits exposed to the atmosphere might find that they are already too swollen to be used. When I put such a biscuit into my microwave oven for a few seconds, it lost almost 10% of its weight – that's a lot of water.

7.11   DENTAL ADHESIVES

There are many specialist types of adhesives used across dentistry. Here we can discuss two very different types.

The first is the denture adhesives that have the difficult task of sticking to the wet mucous surface of the inside of the mouth. In fact, it isn't so hard to stick to the inside of the mouth – the problem is to stick hard when required and, at the same time, be easy to remove when desired. Even this isn't difficult in its basic form. We know that strong butt-joint adhesion is relatively easy to achieve if we have good contact with both surfaces; we need a soft adhesive that flows nicely into place. This strong adhesion is easily countered by peel forces, so the denture, with the right pull in the right place, can also be removed on demand. The problem is that the shear strength of such a soft system is poor and any unexpected combination of forces when chewing some food can result in a shift of position of the dentures which will, in turn, move them out of good contact, allowing easy release (Figure 7.8).

A scientific solution is to use a harder/stronger adhesive that is perfectly moulded to the shape of the denture and, therefore, to the mouth. Stefan's squeeze law tells us that this is not going to

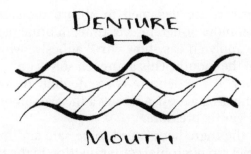

Figure 7.8 At the start, the denture will exactly follow the contours of the mouth. But over time the mouth shifts and the denture can no longer provide a perfect fit with the thin layer of adhesive in between.

be practical with a typical paste adhesive applied with variable thickness. A pre-prepared strip of adhesive does do this and achieves excellent adhesion – for a the first few months of use. Over time, the mouth and/or the dentures change shape, and the trick of using a precision thickness of adhesive no longer works. Some users decide that it is cheaper to accept the limitations of a paste adhesive than to get a new set of dentures with a more perfect fit.

The result of these tensions is that most denture adhesives need to be as strong as possible while still remaining soft enough to conform to the mouth. A good way to achieve this compromise is to use an aggressive form of pressure sensitive adhesive design, as discussed in the PSA chapter. This solves the previous problems – and creates a fresh one. A really strong PSA will fail by cohesion, i.e. it will leave a layer of adhesive on both surfaces. Users really don't like having the inside of their mouths coated with a difficult-to-remove adhesive, so they tend to choose the less good adhesives that fail more easily, while leaving no layer of adhesive on the surface of the mouth.

The adhesive also has an ambiguous relationship with water. It must be happy to accommodate a bit of water, otherwise the wet surface of the mouth will result in a thin water film between adhesive and mouth, rendering the adhesive useless. It must, though, not accept too much water, otherwise it will gradually soften by water absorption during the day. This need for water resistance makes the previous problem of a super-sticky residue harder to resolve, as you can't just rinse away the residue.

Some adhesives are designed to absorb moisture from the mouth and get more adherent over time; in principle this makes it possible to apply an easy-flow "dry" adhesive which becomes harder within the mouth. This approach solves one problem and creates another. If the denture is accidentally dislodged via an unexpected shearing action, a harder adhesive can be less able to re-seal under gentle pressure.

Finally, all the ingredients need to be safe and free from any tendency to pick up microbial contamination in the tube. It takes just one false news rumour on the internet about some chemical being unsafe and a whole, trusted brand of denture adhesive becomes unsellable.

Modern technology has made it possible to design super-accurate dentures and super-capable adhesives. Most of the time, the combination works well. The real challenges are not so much the dentures and the adhesives. Instead, the challenges come from those users who end up battling the laws of adhesion science via ill-fitting dentures and mis-application of the adhesives. Adhesion science can do a lot to accommodate such users, but it cannot achieve the impossible.

The second common type of adhesive in dentistry is the one used for filling cavities. As mercury amalgam fillings have become ever less popular, the technology for creating UV cured (or "blue light") adhesives has advanced to overcome many of the problems of the earlier versions.

The basics are simple: a mix of relatively safe 2-, 3-, 4-functional methacrylates (slower to cure than acrylates but perceived as being safer for oral use), a light-responsive re-action initiator ("photocatalyst", "photocuring agent" or just "photoinitiator"), and some solid particles, such as silica (sand) to provide the hardness equivalent to the solid particles in teeth which are formed from a mineral known as hydroxyapatite.

Real teeth have a design luxury unavailable to simple fillings – they are multi-layered. The outside of a tooth, the enamel, is almost pure hydroxyapatite which is a super-hard mineral. It has two problems

1. It easily dissolves in the acids our teeth encounter from things like fizzy drinks and orange juice or the acid from

bacteria that feed off the sugars from the drinks and our food.
2. Although it is hard, it is far too brittle to last for years of hard use.

Nature has not found a good solution to the first problem. To solve the second problem, underneath the enamel is dentine, a composite of hydroxyapatite and a flexible adhesive system of (mostly) collagen. The dentine is too weak for the surface of the tooth; instead it provides the shock absorbency required to stop the surface enamel from cracking.

A dental filling has to somehow be flexible enough to be tough while hard enough not to erode easily. The earlier versions of the UV systems found this difficult to achieve. Scientifically it was known that lots of small (nanosized) particles would be ideal, yet these consistently failed to provide the required hardness. They therefore added some rather large particles of quartz which provided a sort of brute-force hardness at the surface, even though they were generally unsatisfactory within the filling.

The key to modern versions was the realization that a particle is a source of crack weakness unless it is properly integrated into the cured matrix. By adding special adhesion promoter groups that at one end reacted to the silica and had an acrylate group at the other, the UV curing could now also establish a permanent set of links between the silica nanoparticles and the crosslinked acrylate matrix (Figure 7.9). As a bonus, these smaller particles scattered far less light, so it was possible for the UV light to penetrate deep into the filling to provide cure right the way through.

There is always a downside to any upside. A beautifully transparent filling is optimal for all the mechanical aspects of a

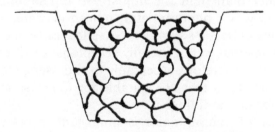

Figure 7.9 The secret to a strong filling is that the nanoparticles are locked into the already hardened UV methacrylate matrix.

filling – and fails in the key aesthetic aspect of being white. Adding particles such as TiO_2 (the super-white pigment in paint) provides the desired whiteness. The whiteness comes from scattering the light back up to the surface, and it is that scattering which stops the UV from being able to create a deep cure. It is therefore necessary to create multi-step fillings in deep cavities. This, on the other hand, offers a fresh opportunity: if patients get used to the hassle of multi-step fillings then the first components can be more flexible, like dentine, and the surface components can be harder, like enamel.

How does the cured system stick to the tooth itself? It turns out that including a modest proportion of –OH groups in the formulation is sufficient to gain compatibility and adhesion to the dentine. Within the formulation a molecule such as HEMA (hydroxyethylmethacrylate) has an –OH at one end and a methacrylate at the other, ensuring that the bonding agent is integrated with the cured resin.

The choice of methacrylates and their functionalities involves the usual trade-offs between hardness and flexibility. Many formulations include a polymer such as PMMA to give a general toughness to the whole system. It takes a lot of work to find the right balance of functionalities of the acrylates in scientific terms.

Then there are the safety issues surrounding acrylates and methacrylates in general. Understandably, there are strict regulatory issues around the types of methacrylates that can be used for dental adhesives. A starting point is the so-called Nestlé list (almost no one calls it by its official standard name which is a Swiss ordnance number). This list of "OK to use" acrylates came out of a media storm around a perceived acrylate safety issue in printed food packaging that somehow became associated with Nestlé. Rather than fight the unjustified reputational damage (the problem was a photoinitiator, not an acrylate, and it was caused by one of Nestlé's suppliers), Nestlé set to work to create a list of acrylates that were generally acceptable to all relevant expert bodies and, therefore, acceptable to end users. Methacrylate versions of acrylates that are suitable for food use are off to a good start for dental fillings.

The choice of photoinitiator is important in both scientific and aesthetic terms. Scientifically, it is necessary to match the absorption of the UV light so that the maximum number of photons get translated into reactive radicals. Anything which absorbs in the UV/blue region takes on a yellow colour, which means that an uncured filling material is bright yellow. Some photoinitiators don't change their colour during the process, so cannot be used, for the obvious reason that people don't want yellow fillings. Others create their radicals by destroying themselves and their colour. It is these types that are used for dentistry.

Finally, there is a choice about the light source. A real UV light is arguably the best for the fullest cure because there are excellent packages of efficient initiators that work well, with each sub-component providing a desirable attribute. A photoinitiator system for blue light is more difficult to find, but such lights are safer to use than UV and have become the default for dentistry.

How long does it take to develop a safe, efficient, effective dental filling material? I remember discussions with pioneers of the technology 40 years ago. They had rather good systems back then, because it is not so hard to make a basic system of acrylates and fillers. As the timescale shows, "rather good" is not the same as "so good that everyone uses them", because even now the technology is evolving to give an even better balance of properties.

One other dental adhesion issue is discussed in the bio-adhesives chapter. This is the amazingly (distressingly) well-adhered dental plaque. The key questions to be answered there are how does the plaque produce such strong adhesion and why is it so difficult to get rid of the plaque entirely?

7.12   FIXING BONES

When we break a bone, it is sometimes possible to set it back into its original position and leave nature to grow the bone back together. Many times, the fracture is sufficiently complex to require re-assembly by artificial means. Generally, this means metal plates, nuts and bolts, screws and other unnatural types of fixing. These work splendidly, but are not going to disappear when their job is done. It has generally been assumed that an adhesive to do a proper fixing job would be available and that surgeons would be delighted to abandon the metal fixings and, like everyone else, glue things together whenever possible.

It hasn't worked out like that. There are plenty of semi-natural adhesives that can be used around fractures to do a job such as sealing a crack. An example is to use the fibrinogen system that is the basis of blood clotting. Mixing a few natural proteins together creates a liquid adhesive which becomes impressively solid impressively quickly. Over time it will dissolve away. Perfect, except that like so many natural systems it's not really strong/tough enough. It simply cannot be used reliably to fix a major breakage in a bone. Alternatives are inspired by mussel adhesives (discussed in the final chapter), sandcastle adhesives (from the worm *Phragmatopoma calfornica*, that can glue sand particles together), or frog adhesives (from *Notaden bennetti*, the crucifix toad). These are all splendid adhesives that are closer to PSAs than to the strong adhesives required, so despite the numerous papers written on them, using bone fixing as an excuse, they have never really stood a chance of being a success.

There are plenty of classic adhesives that do a good job at providing the requisite strength and reliability in a complex environment; superglue isn't bad at all for fixing some types of fractures. Unfortunately, the effective adhesives are non-biodegradable and some of them have shown sufficient side effects to be unacceptable. It is not surprising that the good adhesives are non-biodegradable. If water and enzymes could easily get into them to start the process of degrading, then they would not be strong enough to be useful.

At the time of writing, the most up-to-date review that I have been able to find paints a glowing vision of adhesives that could potentially be used, while concluding that none of the many systems that have been tried is up to the job and, worse, that there is nothing on the horizon with sufficient early evidence and encouraging preliminary trials in humans to suggest that a solution will appear any time soon.

I had not expected this. I had looked forward to describing the smart science (maybe even bio-inspired) that solved this tricky problem. It was not to be.

7.13   TEMPORARY ADHESION

Finally, we look at systems designed for the non-trivial task of providing the right amount of temporary adhesion.

One approach which we have discussed in Chapter 3 is via a gecko-like method where the material is sufficiently conformal, and the surface sufficiently smooth for pure surface energy to provide the required adhesion. We know that this can be strong in butt-pull mode, adequate in shear and desirably weak in peel mode, making it easy to remove.

Assuming that we don't have a surface smooth enough or clean enough for pure surface energy to give us the desired level of adhesion, getting temporary adhesion first requires a soft material that can flow or be pushed into the rough surface to gain enough surface contact for adhesion. That is easy. The material must also be 100% tangled with itself so that when pulled it all comes off together, leaving no residue behind. That is also easy. The hard thing is to combine both properties in the same material, along with enough dissipation to absorb crack energy at the interface.

Take the familiar (usually blue) putty used to stick posters to walls. A sticky rubber (one component) on its own would provide great adhesion (it isn't too far from the solvent-based real adhesives discussed earlier) but would be nearly impossible to remove as it would stretch and break, like chewing gum. The other key component, chalk, can be made into a cohesive ball, with no adhesion. A combination of the two can provide just the right balance of stickiness and cohesion. If any of us were given a lab with a few samples of soft rubber materials and some chalk, we could probably create a plausible version quite quickly. The hard part for something that claims to be removable is the long-term testing on a variety of good and bad surfaces to ensure that:

- The putty doesn't get so brittle that it falls off over time
- The putty doesn't get so adhesive over time that it pulls the paint off the wall
- Colours and any oils don't leach out, leaving a stain on removal

As with most adhesives, getting 90% of the way there is easy. The final 10% is what separates a successful from a failed product.

A different type of temporary adhesion is one on which our lives depend most days – adhesion of tyres to the road. At the basic level this isn't at all hard. Get some natural rubber that is naturally sticky (via a PSA-style dissipation of energy) and harden it up enough (crosslinking with sulfur, "vulcanization") so it is not too squishy. Form it into a tyre with some woven fabric, using heat and pressure to define the shape. On cooling there's the tyre.

However, our problems have only just begun. If the tyre is smooth and soft, the adhesion will be wonderful in the dry and useless in the wet (because water gets trapped between the tyre and the road). The tyre will also wear out after a relatively short distance and prove to be energy inefficient because of its high rolling resistance (the dissipation that gave the adhesion). If the tyre material is super hard, adhesion will be less (bad), lifetime will be longer (good) and rolling resistance less (good for efficiency). Treads help expel the water for wet grip, while reducing dry grip. For tyres, everything is a compromise.

Because no single material can meet all the requirements, tyres are complex multi-layer materials, with each layer helping with a key property such as resistance to the forces during cornering. Some components are added to provide resistance to the UV of sunlight. A tyre with insufficient protective chemicals will deteriorate alarmingly if parked all day in the sunshine. The tyre mustn't get too soft on a hot day, yet must not be too hard on a cold day, so again there are various components and additives that help keep the properties adequate over a wide temperature range. These multiple components are appreciated by all of us because our tyres perform flawlessly for many miles. However, we get upset at the end of life of the tyre because these multiple components are very difficult to dispose of in an environmentally benign manner.

We also get upset in contradictory ways about the rubber in tyres. It sounds nice to make them from a natural resource, until we see the vast areas of rubber plantation monocultures required to supply the rubber. Swapping to other natural rubbers such as guayule (a scrubby desert plant) shifts the problem to a different type of monoculture. We can avoid these plantations (and the pesticides and fertilizers they require) by making artificial rubbers efficiently from petrochemicals, but these are criticized for being non-renewable.

Temporary adhesion of a tyre to a road is relatively simple in theory and full of tough compromises in practice. In this case adhesion is a property of a complex system of materials in the tyre and a complex set of compromises in providing the materials to build them and in disposing them once the tyre becomes unsafe to use.

CHAPTER 8

Watching Paint Dry

We all want paint that is easy to apply to any surface and will stick forever. At this stage of the book we know how to achieve *most* of that, and why it is impossible to achieve *all* of it. What is amazing is that modern paints are closer to our ideal then they have ever been. Because I am bad at painting, my family has relied on a painter who has looked after our house for decades. He is very much *not* one who believes in the "good old days". For him, modern paints are astoundingly good compared to when he was a young apprentice. What might have taken a few coats, lots of preparation and exposure to lots of fumes can now be done with a single coat of a safe, water-based paint that will last for years.

8.1 A GOOD BASIC PAINT

In the bad old days, the best basic paints were solvent based – for a good reason. A key requirement for any paint is to resist water, and you are off to a good start if your whole paint formulation is insoluble in water and soluble in solvents. Safety regulations, limitations on VOC (Volatile Organic Compound) emissions, and odour issues have pushed the industry towards water-based paints for home use. For *technical* paints where performance is critical and the other issues can be handled professionally,

Sticking Together: The Science of Adhesion
By Steven Abbott
© Steven Abbott 2020
Published by the Royal Society of Chemistry, www.rsc.org

solvent-based paints are still common because they can be formulated to do almost anything. However, we will continue here with just the water-based paints. For the moment we will put to one side a rather important set of ingredients – the pigments that provide the colour.

Because even water-based paints must be water resistant, the trick is to provide a water-insoluble polymer in the form of soft, small polymerized particles suspended in the water, i.e. as an emulsion. Each particle has surfactant molecules on the outside that repel other similar particles, either because the particles are electrically charged, and like charges repel ("charge-charge repulsion"), or because they have surfactant chains that stick out and don't like to get in each other's way. These surfactants can also take on the role of creating the particles in the first place because, in general, paints are created via "emulsion polymerization", where the starting point is a surfactant-based emulsion of water-insoluble monomer drops that become polymerized into the soft particles required for the paint.

Getting this polymer emulsion onto the surface with a brush involves many complex processes which we will ignore for the moment. The essential point is that we want to have a thin, fairly even coating of the emulsion on our door or wall and we need it to become water-resistant rather rapidly, whilst at the same time maintaining perfect molecular contact with the surface to provide the basic minimum of adhesion. As the water evaporates, the polymer particles move closer together, still staying in contact with the surface. As the water finally disappears, the particles have a choice:

1. They can remain isolated and we have a weak layer of polymer particles which are not a good paint.
2. They can coalesce to form a single coherent film.

If you read the labels on paint cans you will see that they often have a small % of solvents other than water. Their main task is to be "coalescence aids" (we came across these in Chapter 5 as the cause of the odour for PVA wood glues) that soften the polymer sufficiently for the individual particles to meld together. These solvents are designed to be relatively slow in evaporating (which is why we need to allow the paint plenty of time to dry), to allow full coalescence before the aids, too, evaporate, leaving a uni-form surface of tough, water-resistant polymer (Figure 8.1). They are especially difficult to get right for indoor use because not only must they be superb at their coalescence job at the lowest-possible levels (manufacturers like to tell you that they are super-low in VOCs), they must also have very low odour, and evaporate slowly enough to do their job, yet fast enough to ensure that any residual odour does not annoy the user for too long.

If you need to apply a second coat it is a good idea to do it while the first coat has not quite fully coalesced because then the particles of the second coat have a chance to meld with those in the first. This advice is even more necessary for the "no VOC" paints which have managed to replace these volatile coalescence aids with non-volatile equivalents which have the obvious drawback of hanging around in the formulation. They can mi-grate to the surface and cause problems with subsequent layers.

A paint made via these emulsion films would show sur-prisingly good adhesion. If you stick some tape to the surface and try to pull it off, that is equivalent to a butt joint pull. There is no edge to create a crack, and the tape will come off before the paint. Rubbing would be a little like a shear force; again, there is no reason why it would fail. As a thin coating it is also difficult to find any way to "pick" at an edge to start peeling it off.

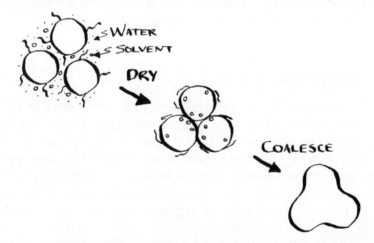

Figure 8.1   As the water dries, the coalescence solvent concentration increases sufficiently to make the particles fluid enough to coalesce, forming a tough, continuous film.

Nevertheless, it will peel off easily on any exposed edge, so we have to find some extra adhesion to create a real paint.

We know how to do this. The paint film is tough and flexible, so a few links between the paint and the surface will generate a lot of entangled dissipation across many nm, and the paint will remain adhered. Many paints contain small amounts of acrylic acid groups built into the chain; these provide some of the water compatibility and charge-charge repulsion to keep the emulsion stable. They are also able to interact reasonably well with chemical groups on the surfaces of bricks, plaster, metals and wood, producing the modest levels of chemical interaction we need at the interface to provide adequate adhesion. For coatings onto previous layers of paint, the coalescing solvents can usually provide enough "bite" into the previous layer to allow entanglement between layers. This will not happen on old and brittle paint, which is why removal of such old paints is highly advisable.

For added toughness, reactive polymeric systems are required. The classic system is based on polyesters known as alkyd resins that provide the "backbone", while side chains of natural oils such as linseed or sunflower provide reactive double bonds. As the paint dries, oxygen in the air initiates reactions with those double bonds, and the paint crosslinks. As we have seen many times, the level of crosslinking has to be carefully controlled because if the crosslinking is too severe, the paint becomes too brittle.

8.2   PAINT PIGMENTS

There are whole books written on the colour aspects of paint pigments. The first time I became engaged in some colour science, a colleague and I agreed that we had better spend a few hours reading up on the theory so we could solve some problems. How naïve we were! Colour science is fiendishly complicated. Fortunately, it is not our concern in this book. Let us just accept that we have plenty of pigment. What are the adhesion issues?

I was once asked to help solve a difficult paint adhesion issue. The paint (it was an ink rather than a paint, but the principles are the same) was losing adhesion during tests that involved rubbing it with some specific chemicals. I spent time thinking of possible causes and cures, all of which were perfectly sens-. ible. Finally, I got hold of some test samples to see the adhesion failure for myself. I could immediately throw away all my ideas. The paint was adhering perfectly. The rub test was dragging out some of the pigment particles. The adhesion problem, therefore, was not of paint to the substrate but pigment to the paint.

This story has three take-home messages:

1. Never try to solve a problem until you know what the problem really is
2. Pay as much attention to adhesion *within* a paint as to adhesion *of* the paint
3. Whenever possible, turn a problem into an opportunity

Pigments have two problems, and therefore two opportunities. The first problem is that in the can of paint they must remain perfectly separated from each other and from the emulsion particles. The solution is to cover them with dispersants, typically surfactant-like molecules, just like the emulsion particles. The second problem is that the pigments themselves are generally designed to be as unreactive as possible, otherwise over time (and especially in the UV rays from sunlight) they would start reacting with, and destroying, the polymers in the paints. This means that it is not possible to directly react the pigments themselves into the matrix.

The opportunity is to provide dispersant molecules that are strongly locked to the pigment surface (this can be done at an early stage in preparation of the pigments) and where the other end of the dispersant can entangle into the polymer matrix of the paint via some sort of reaction (Figure 8.2). In an alkyd paint where crosslinking is via double bonds, a dispersant which is locked onto the pigment particles at one end should have double bonds which will be able to react into overall paint system. Not only is the pigment now bonded into the paint, the paint itself is "bonded" to the pigments. The pigments are strong, so what was previously merely "colour" and the source of a potential problem is now a core part of the overall paint integrity. This was the solution to the "paint adhesion" problem mentioned above.

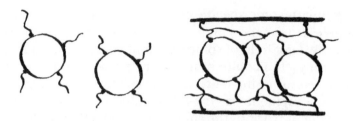

Figure 8.2   The polymers that keep the pigment dispersed can be chosen to react as the paint dries, locking the pigments securely into the paint structure.

The pigment has one further important functionality. Most practical polymers used in paints will be damaged by UV radiation from sunlight. In addition to absorbing and scattering specific visible wavelengths of light to generate the desired colour, the pigments absorb, dissipate and scatter UV energy, reducing the damage to the polymers. So, by primarily acting as colour filters, pigments add greatly to the longevity of the paint.

There is an interesting proof of this. To explain, I must go back to when I was senior enough to be in charge of a coating process and junior enough to have many gaps in my basic knowledge. We started the coating machine and everything seemed fine, until we looked at the final, dried coating and found that the whole surface was covered in hexagons. Nothing in the machine was hexagonal, so where did the hexagons come from? Because I had no idea, I went to see our science guru. As soon as I said "hexagon" he said: "the Marangoni effect" and from that point on, the problem was easily solved.

The Marangoni effect (Figure 8.3) starts when any part of the surface of a paint or coating has a different surface tension thanks to a slight difference in temperature or composition. The paint flows from low to high surface tension, creating a mini wave on the surface. The same thing happens on different parts of the surface and various paint flows are set up. Some of them reinforce each other, others cancel themselves out. After a short time, the reinforcing flows win out. The only way you can get a set of self-reinforcing flows is if they create a pattern of hexagons. The principle is general; if you carefully heat a pan of water and add small amounts of an insoluble fine powder as a marker you can see a pattern of hexagons appear on the surface.

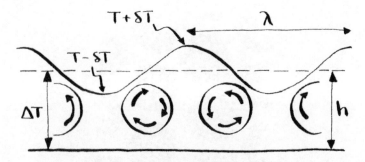

Figure 8.3 The Marangoni effect spontaneously creates hexagonal structures created by surface tension differences that make the fluid flow.

In the case of my coatings, we had a mixture of solvents with different surface tensions; one evaporated much faster than the other, creating big surface tension gradients that generated the cells. By swapping to a similar solvent that evaporated less quickly, the problem disappeared.

In some applications, where designers deliberately want a nice "hammered" appearance to the paint, the Marangoni effect with small cell sizes can be encouraged by careful selection of solvents, viscosities and paint thicknesses. In water-based paints, the same effect can be created via subtle surfactant effects.

But the question that started this detour into Marangoni was: are we sure that pigments really protect our paints? Several decades ago, some ingenious experiments were carried out to coat the same substrate with the same paint while drying under conditions that deliberately encouraged or discouraged the Marangoni effect. With the right choice of pigments in the paint it was especially easy to see the Marangoni cells because the pigment particles were dragged from the cell walls which, therefore, were devoid of colour. The samples were subjected to ageing tests that simulated strong sunshine, hot temperatures and plenty of rain. The paints without the Marangoni cells hardly noticed the tests. Those with the cells showed strong failure at the cell walls because the pigment that should have been protecting against the UV rays was no longer present (Figure 8.4).

This was a long diversion to make a straightforward point. I find it so fascinating that hexagonal patterns can appear spontaneously that I like to find an excuse to talk about the Marangoni effect. Of course, the research showing that

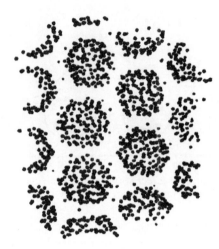

Figure 8.4   The Marangoni flows in a paint can strip the pigment particles from the hexagonal cell walls, leaving pigment-free paint that is easily destroyed by light, temperature and humidity.

Marangoni not only causes uneven surfaces but also can lead to long-term failure in the paint forced the paint industry to think hard about solvent blends and surfactant systems. The result is that most paint formulations (except those designed for a hammer finish) are now very far from any risk of Marangoni effects.

8.3   WOOD PAINT CHALLENGES

As mentioned at the start of the chapter, I am no good at household painting. I lack the skill, knowledge and patience to do a good job: we call in the professionals for that. But when, as on one occasion, a professional needed a low-level assistant in tackling a large job in our house, it turns out I could help – and learnt a lot in doing so.

The key lesson was that wood primer paints really are a good idea and really do save both time and money. The primer can be slapped on quickly even by someone as unskilled as me, and, by not having to act as a paint alone, solves at least two problems.

1. Wood and plasterboard (the job involved both) are absorbent. The primer can put enough pigment and binder quickly and easily *into* the structure so that the top coat can remain *on the top* and do its decorative job, rather than disappear inside the pores of the substrate.

2. It is hard to get good adhesion on, into and around the fibres in the wood or board. Although the top coat *can* do this, the primer can do it better as it does not have to worry about looking nice. The primer can contain more binder than pigment in its formulation, resulting in more binding. It can also contain binders that are more compliant towards the shifts in the fibres, allowing a greater resilience to the overall system.

As long as you don't wait too long between priming and top coating, the chemistry in the top coat can easily entangle with the primer chemistry so that the whole system is more secure (Figure 8.5).

For indoors painting (which is all I was allowed to do), that is almost the end of the story. Any reasonably tough paint can withstand the general bumps and scratches from daily life by being neither too soft as to be scratchable itself, nor too hard that it would crack when the softer surface (wood or board) bends beneath it.

The challenges for wood paints are more severe in outdoor environments, though not for obvious reasons. The problem of the UV light has already been discussed; modern pigment/binder combinations really have little problem surviving most solar environments. Making the paint resist downpours of rain is also

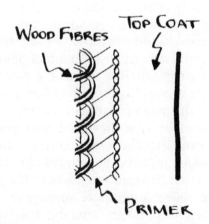

Figure 8.5 The primer can wrap readily around wood (or plasterboard) fibres and also entangle with the topcoat for strong overall adhesion.

not a problem – the water bounces off a basically non-absorbent paint surface. The real problem is the weather.

If you tell a formulator "This paint/coating must withstand high temperatures" they'll say; "Fine, no problem". They will say the same if you mention high humidity, on its own, or lots of UV. What strikes fear into any formulator is the requirement to deal with combined cycles of heat, humidity and UV. As we know, what kills adhesion is stresses. An ordinary paint in a stable environment quickly equilibrates to a relaxed state. With no stresses, there is no failure. A change of humidity will change the stresses in some parts: the wood will certainly expand, and the various coating layers, primer and top coat may or may not absorb moisture to varying degrees. UV may attack some chemical bonds, which is not a good thing but isn't a big deal unless water comes along and washes out some of the newly-created smaller molecules. Now there is less coating over the same area, so there is a tendency for it to want to shrink.

Any one of these stresses can probably be survived. What is hard to survive is the constant changes of these stresses.

Some obvious solutions have unexpected downsides. If the paint is made super water-resistant it will not expand or contract with humidity. Unfortunately, the wood below *will* expand and contract as water vapour makes its way slowly into the structure. Go the other way and make the paint highly water permeable and absorbent, now the wood receives insufficient protection from the water and can become saturated. At one time there was a craze for "microporous paints" for outdoor wood – the argument being that the wood could "breathe" by permeation of water vapour and not build up stresses. This had always puzzled me because it is almost impossible to make a practical paint that doesn't allow water vapour to permeate. I was, therefore, relieved to find that one of the major paint manufacturers at the time, ICI Paints, refused to claim that they used microporous paint. ICI explained at the time that the term incorrectly implied coatings could be made so that moisture vapour could only travel through them in one direction. Instead scientists at ICI preferred to describe paints and stains in terms of appropriate levels of moisture vapour permeability and outlined that this was only one of a number of important attributes to be considered when formulating exterior coatings including adhesion, cohesion, flexibility, mould resistance and photostability. I found it rather refreshing that a paint company would state the science rather than go along with marketing hype.

There have also been some spectacular wood paint failures when the paint has varied from hard when dry to soft when wet. The water can get through the paint very easily when it is wet. When the top surface dries, it becomes water impermeable, trapping too much water inside. In hot sunshine the water that is still trapped beneath the outer layer can expand and create a blister.

One approach which is of general applicability is to make the primer well suited to absorbing the local stresses of the expanding and contracting wood, with the top coat well suited to surviving the UV and thermal stresses. This means that the primer will be generally less crosslinked than the top coat. If, by clever chemistry, the interface between primer and top coat can be made to transition gradually, that gets close to the important multilayer mechanics approach discussed below.

The testing of outdoor paints is very severe, often requiring several years of exposure to harsh weather in hot, humid, sunny places such as Florida. Once, a set of samples from the company I worked for failed to last the entire time but were still passed as officially OK. The reason for this is that one month before the end of the test period (when they still looked good), the entire test site was destroyed in a hurricane. Special certification was given to those of us who had nearly completed the whole test cycle.

Super-severe tests required special locations near a convenient tropical sea. These provides a high level of salt in a hot and humid atmosphere. A different type of salt test is described shortly.

8.4 AUTOMOBILE PAINTS

These days, painting a car is far easier, yet far more demanding than in the bad old days. I'm from a generation that regarded paint failure on a car and the subsequent rust repair as normal. I now cannot recall when I last saw any rust on a modern car. The transformation started with the invention of electro-deposition of paints, and this made painting far easier and more reliable.

The metal part of the car is thoroughly cleaned then passed into a tank of paint. A voltage is applied to the metal and the oppositely charged paint particles are attracted to every portion of the metal, inside and out. This first coat is full of phosphate groups that love to attach strongly to the metal surface and are excellent inhibitors of rust on steel. Any excess is allowed to drip off, then the painted shell goes through the processes of precision drying, heating and curing.

The next coat is a thick "primer" layer, typically spray coated, which provides a lot of protection, especially providing extra resistance against chipping from, say, flying stones. The upper paint layers have other functions which restrict the ability to make them super-tough. Although these upper layers might get damaged, at least the metal is protected by the primer, so a quick re-touch will make an adequate repair. Because the paint is handled in a purely industrial environment there is a wider choice of chemistry that can be used to provide intermingling, entanglement, chemical bonding and crosslinking – everything required for robust adhesion. An alternative is to apply the paint as a powder which flows and cures when heated. In either case the result is a tough, high quality paint surface.

The main colour coat is next applied in a similar manner.

Then one final coat is required – a transparent gloss coat which provides a further layer of protection.

This whole process is far superior to the old ways of spraying, and far easier – once you have spent the $millions to install the equipment and spent years agreeing deals with your paint supplier, who usually runs the paint shop within the manufacturer's production line. Why, then, is it also more demanding?

Because we are so used to car paint being reliable that we can choose to treat our cars as badly as we wish and expect the finish to remain perfect for as long as we own the car. This is an exceptionally tough challenge, so the industry has to go to great lengths to test for all the things that might go wrong. Before discussing the science of multilayer adhesion as a strategy for toughness, there is an interesting example of failure that led to an industry standard of "Russian mud" (a mix of calcium chloride which is an anti-freeze salt and kaolin clay which simulates road dirt) used in testing. This standard mud allowed the next generations of cars to survive Russian winters.

The story goes that a major European car manufacturer managed to sell lots of cars to Russia. The cars were fully capable of withstanding European winters, and there were no obvious reasons why they would fail in Russian ones. Yet they failed spectacularly and started to rust. The reason was that European roads were "salted" as an anti-freeze during winter with common salt, sodium chloride. Corrosion tests involved lots of sodium chloride sprays, and the cars passed the tests with no problem. The Russian road salt includes calcium chloride, which had not been included in standard tests and which happens to corrode steel via different mechanisms. When a car got covered in Russian salty mud, corrosion set in. All automotive paint suppliers now have stocks of standard Russian mud to test their latest paint formulations.

8.5   MULTILAYER PAINTS AND COATINGS

The automotive coating is a typical example of a high-performance multilayer system. The simplified description focussed on the main properties of each layer: strong adhesion to the metal; providing a good tough primer coat for general robustness and a layer of protection even if the colour is chipped or scratched off; providing the colour and general toughness; providing a tough gloss coat for the outer layer.

In multilayer coatings there is another key element that yet again emphasizes that adhesion is a property of the system. As a specific example, we can discuss the antireflection coatings on glasses (or spectacles if you prefer that word given that glasses are no longer made from glass).

At one time I worked on a bio-mimetic antireflection coating. Moths need large eyes to see in low light levels, but reflection off these large eyes would easily be visible to predators. The moths therefore developed a nanostructured "stealth" surface with very low reflectivity. We were able to create similar structures with impressive antireflection capabilities – but they were never a success for two reasons:

1. Finger grease got trapped in the nanostructure and destroyed the antireflection capabilities
2. It was difficult to make them strong enough to resist scratching

I became interested in the amazing toughness and scratch resistance of the normal antireflection coatings on glasses. The antireflection system is a series of vacuum-deposited oxides and fluorides that are themselves rather hard and strong. I could see why they would be scratch resistant when deposited onto glass-based glasses. My problem was that everyone wears plastic-based glasses and I knew that the thin, hard coating stood zero chance of a long life if deposited straight onto the plastic. Pressure on the surface would cause the softer plastic to deform, causing the thin rigid coating above it to crack.

The solution to the problem partly explains why these coated glasses are relatively expensive. Between the surface of the polymer lens and the vacuum coated surface are a number (at least three, usually more) of thin polymer coatings. The trick is that the coatings get gradually harder. When something tries to scratch the top layer, the layer deforms into the layer below; because that lower layer is not too soft, the deformation is not too large so there is no build-up of sharp stresses that could crack the coating. The same thing happens at each interface – there are no reasons for large stress concentrations. The result is that the energy that might have scratched the surface is dissipated by each layer absorbing just the right amount of energy without generating any serious stress gradients.

The layers of automotive paint are also carefully designed so that in addition to all their other requirements, the sudden forces from, say, a flying stone are never too focussed on any one interface. The coatings together resist the energy of the stone via a chain of influence rather than by trying too hard.

The principle, in reverse, is behind the safety windscreens (windshields) on our cars. They are multiple laminates of strong but easily shattered glass, interspersed with soft but absorbent, non-shattering PVB, polyvinylbutyral. PVB is an excellent adhesive for glass – in that it can be hot laminated into perfect contact with each layer and is tough and flexible without being especially strong. It is not that the PVB is there merely to hold broken pieces of glass together; that is a feature of last resort. If a stone hits the glass, there is a good chance that the forces are fully absorbed, dissipated, by the relatively soft PVB so the stone bounces off harmlessly. When it is a bullet hitting the glass, the PVB isn't expected to make the bullet bounce off; instead the bullet has to expend a lot of energy to break the glass *and* to move enough glass out of the way to get to the PVB through which is can easily pass, then face the same set of challenges with the next layer of glass. The fact that PVB is *not* hard means that the bits of glass, still stuck to the polymer, dissipate the energy by stretching the PVB.

8.6   POWDER COATINGS

As mentioned, one way to apply automotive paint is via a powder coating process. The same process is widely available across many industries because it allows a thick, tough coating to be applied without the problems of having to remove the solvents or water needed in conventional paints.

A powder is electrosprayed onto a grounded metal part. The thick layer (powder coating is bad at giving thin coatings) self-adheres via the static charge. The part is then heated to start sintering the paint (sintering was mentioned in Chapter 7 and is discussed in detail in the next chapter on 3D adhesion) then heated further to get the paint to flow. Thermal curing reactions via urethane, acrylates or epoxies kick in slowly enough to allow the paint to flow and form the required coating, and fast enough for the coating to be cured by the end of the process.

Surface preparation and the provision of suitable primers has to be especially good because getting perfect flow and wetting at the interface is much harder than with conventional paint. The shrinkage during the high temperature cure and subsequent cooling will provide some extra security to the coating which, being thick, is in any case providing self-protection. I admit that there is a contradiction here to the general rule that shrinkage adds unwelcome stress and thickness causes problems. Because a thin powder coat is not practical to apply, formulators have had to learn how to get the benefits from shrinkage and thickness without suffering too much from the downsides.

An alternative powder coating of increasing popularity is a UV acrylate system. It is delivered, of course, as a powder, but it requires only a burst of surface heating to melt it, then a short burst of UV light to cure it. This consumes only a small fraction of the energy needed to fully heat and cure a conventional system. Because the UV curing is done on the hot acrylate, the degree of cure is higher because there is more freedom of movement within the curing system, as discussed in Chapter 5.

8.7   VISCOSITY: THICK AND THIN

It is delightfully easy to paint with a "thin" (low viscosity) paint. The brush glides along easily and the brush marks (discussed in the next section) disappear quickly. However, the downsides of a thin paint are equally clear. Getting the paint from the can to the surface results in a lot of splatter on the floor, the paint flows down the bristles onto the handle when you are reaching up high, and the paint quickly beads up and runs down and around any edge or corner, leaving an unsightly dribble mark.

A thick (high viscosity) paint is the other way around. Everything about it is much easier except for the painting itself which is hard work (including the permanent brush marks).

What we all want is an intelligent paint that is thick during all times other than when the brush strokes are being applied, when we want it to be thin, and remain thin long enough for the brush marks to flow out.

We already know from adhesion science one way to do this. Tangles are a wonderful way to resist forces, so an entangled paint is nice and thick. If we have the level of entanglement just right, then the differential motion created by the brush strokes will stretch out the polymers parallel to each other, reducing the amount of entanglement. The technical term for the differential motion is "shear". Processes other than brushing are "low shear" and brushing is "high shear".

If you take a range of different polymers, each in a good solvent and each with the same molecular weight so we are comparing like with like, we can measure the viscosity at low shear and at high shear. All of them will have a lower viscosity at high shear, because the polymers stretch out and provide less tangled resistance to flow. For some the reduction in viscosity will be modest, for others it will be large. This is because some polymers self-tangle much more than others. The transition from high to low viscosity will take place at different shear values because different polymers will require more shear to disentangle.

With solvent-based paints it was possible to find adequate combinations of polymers that could achieve some of this balance. For water-based paints a different type of science comes into play because we no longer have our polymers dissolved in the solvent – they are there as emulsion particles. At low shears the individual emulsion particles get in each other's way, just like cars trying to thread through a crowded street with local traffic going in all directions. At high shears, the particles line up, like cars on a high-speed road, and can move along with far less resistance. This means that just about any emulsion paint will show some reduction in viscosity when sheared – the desirable behaviour comes automatically with the physics of flow.

In many cases, though, the automatic effect isn't strong enough to give us the performance that we want. To get a stronger difference requires entanglement of a different kind. We saw earlier that each emulsion particle has to be protected from the other particles by a dispersing agent – otherwise the whole can of paint would coagulate to a single lump. If we make those agents a bit smarter then we can arrange for them to create an entangled network. One possibility would be to have them as long polymer chains. The problem is that the resulting transition from high to low viscosity would be rather slow and tricky. Instead the molecules are arranged to form lots of relatively weak, temporary bonds such as hydrogen bonds between the H of an –OH group and the O of a different –OH group. This is especially appropriate in water-based systems because water itself creates an extensive network of such hydrogen bonds. We end up with a kind of competition between different bonds and repulsions; water–water, water-dispersant, dispersant-dispersant, so that the forces can be nicely balanced to give the high viscosity we want at low shear and equally, because the weak bonds can be ripped apart, low viscosity at high shear.

Paints that can change their viscosities when sheared are often called "thixotropic". What I have described so far is called "shear thinning" or "pseudoplastic" behaviour. Thixotropy is different. Yes, a thixotropic material undergoes shear thinning. The difference is that it then takes time to recover to its original viscosity. Some thixotropic paints are designed to be too viscous in the can to be conveniently put on to the brush – you first have to stir them vigorously so they shear thin *and remain thin for a significant time*. The trick when designing for thixotropy is knowing how long the paint should remain thin. If it goes thick immediately after applying with the brush, there will be no chance for the brush marks to flow out. If it stays thin too long, then it can start to dribble down a vertical surface. Control of the timing is based on the number of self-associating groups and how freely the polymer chains to which they are attached are able to move (Figure 8.6).

We are left with a final problem. All those water-loving, hydrogen-bonding groups are wonderful while the paint is liquid and a potential source of problems when the paint has dried – we really do not want water to be attracted into the paint where it can cause problems.

The solution is to ensure that as the emulsion particles fuse together in the presence of the coalescing solvents, the hydrogen-bonded network self assembles as strongly as possible (adding extra crosslinks to the system) so that dispersant-dispersant interactions resist the potential interactions from water that gets into the system. These strong interactions also make the dispersant molecule network more compatible with the water-hating polymer system, helping the network to more

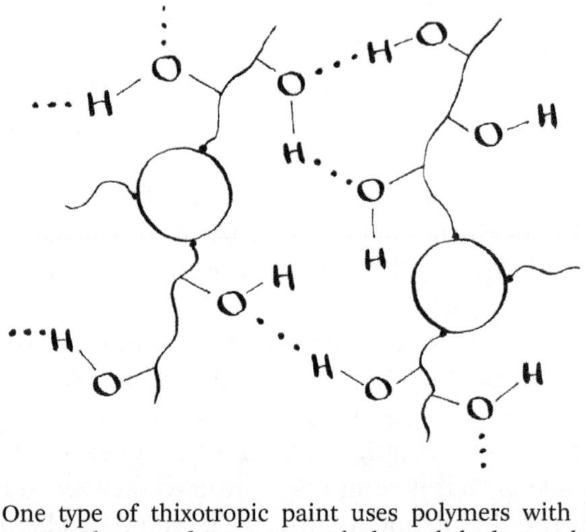

Figure 8.6 One type of thixotropic paint uses polymers with lots of –OH groups that can form a network through hydrogen bonds to give a thick paint. The network is easily broken with shear and re-forms after a time that can be tuned via design of the polymer.

easily end up inside the polymer rather than in clumps where the edges of the particles used to be.

There are some final tricks to paint and painting with which to end this chapter.

8.8 LOW PRICE, HIGH COST

For many years I fought the laws of physics when painting, without realizing that I was losing every time. To me, a paint brush was a paint brush. Why spend twice as much on a fancy brush when a cheap one will do the same thing – pick the paint out of the can and slap it onto the surface?

Everything changed when I had to study the science of how paints and other coatings "level", i.e. if you have some regular pattern such as brush marks, how quickly do they flow out to give a smooth coating? As soon as I found the formula, I realized that I had been wasting money by buying cheap brushes. I was also giving myself the problem of paint "slump" on vertical surfaces such as doors – where an over-thick layer just sags and gives a thick, uneven layer near the bottom.

We can imagine that our brush marks have a regular spacing between them of λ and that the thickness of the paint (not the thickness of the brush marks) is h. The theory (described in an app: https://www.stevenabbott.co.uk/practical-coatings/levelling. php) tells us that the time for the marks to level out depends on λ^4/h^3 (Figure 8.7).

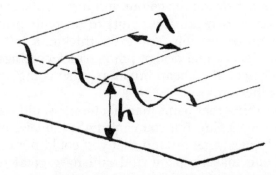

Figure 8.7 Your cheap paint brush will produce lines with a wide spacing, λ, that will take much longer to flow out than the fine lines from an expensive brush. The only fix is to increase the thickness of paint, h.

If I have a cheap brush with course bristles and an expensive one with fine bristles, λ for the cheap brush will be, say, twice that of the expensive brush. The equation tells us that the time for my brush marks to level out will be $16\times$ longer! The only thing I can do to help the marks disappear is to double the thickness of my paint, h, which reduces the levelling time by a factor of 8, so I am only a factor of 2 slower which is probably OK. Now I have two problems.

1. I use up twice as much paint, costing me more than the extra price of a better brush.
2. I increase the risk of a different major visible defect, slump, because the tendency for paint to flow into a blob increases as h^3. In fact, the tendency to slump is worse than h^3 because thicker coatings take longer to dry, and their viscosity stays lower for a longer time, allowing more time for slump to occur.

Does this have anything to do with the science of adhesion? Well, yes. Our instinct is that thicker coatings should give better adhesion. In terms of resistance to UV that might be true. For everything else thickness is, on balance, a bad thing. The differential stresses in a thick coating are higher, the chances of accidentally chipping away a thicker coating via an exposed edge are higher, and the chances of trapping a water layer, or absorbing a large amount of water are also higher. Paints have been designed to be optimal when there is enough thickness for the pigments to provide the colour and hiding power (ability to hide whatever is beneath the paint) to do their primary decorative and protective job. Any extra thickness is taking the paint beyond its optimal design point. Thick coatings to make up for the defects of a cheap brush are also going against the physics of adhesion.

You might think that paint manufacturers would want you to use a lot of their paint. But you can't fight the laws of physics. Word would quickly get around that you could put on half the amount of one maker's paint and still have great results, so everyone has to follow. This isn't a theoretical point. In the bad old days, paint technology was sufficiently crude that you had to have two thicker coats to be able to get the hiding power and general resistance to weather and day-to-day damage. These days we expect a one-coat paint to be so good that you can even get away with not using a primer (though the results will not be as good), so, effectively, we have all arrived at half the coating thickness.

The ability to create an effective one-coat system has evolved partly through improved pigment technology, partly through new resin technology and, to a large extent, through the application of the core principles of adhesion science. Everything has to work in harmony with the laws of physics and chemistry:

- The pigment dispersants must disperse well and also allow controlled lock-in of the pigments as part of the overall toughness profile.
- Paint to substrate binding is controlled by the right levels of additives that can lock onto the surface and into the overall paint structure.
- Smart phase separations can be designed in to bring some parts of the paint to the surface to provide excellent gloss or a controlled roughness to give, say, a satin finish.
- The top coat should entangle nicely with the primer polymers which in turn are entangled with the surface.
- The water should escape relatively quickly so the paint is touch dry for the impatient user, while the coalescing solvents escape slowly enough to allow the emulsion particles to fuse together to give a resistant coating.
- Any can of water-based polymer is potential food for microorganisms, so the paint should be as unfriendly as possible to them while not containing any chemicals that might scare the public. Similarly, the coalescing solvents should be wonderful at coalescing while being perfectly safe and as odour-free as possible.

I used to be shocked at how expensive a typical can of paint seemed to be. Now, I marvel that so much technology is affordably available and that most of the time most of us can do an adequate job of getting the right colour onto our chosen surface, with adequate visual quality and for it to remain adhered to the surface for years or decades.

8.9   SAVING THE PLANET

It is well-known in the industry that water-based paints are not automatically as planet friendly as they seem. This sounds surprising and non-intuitive. The gold standard for judging the impact on the planet is an LCA, Life Cycle Analysis, which examines every step in the process to check the impact on the planet in terms of CO_2 emissions, waste, Volatile Organic Compounds (and their potential to create ozone and nitrogen oxides), resource utilization, renewability and so forth. The LCA for some water-based paints turned out to be worse than for some classic solvent-based paints. My understanding is that the serious paint manufacturers have learned important lessons from these studies and that the LCA of water-based systems continues to improve. One example of how a solvent-based system might have an advantage involves the whitening pigment, TiO_2. Some solvent-based systems have a trick whereby they create lots of micro-voids, which scatter light effectively. This means that there is less need for energy-hungry TiO_2, leading to lower CO_2 emissions.

Another example of the difficulties of providing a more sustainable paint involved a large European project. The project team accepted that solvent-based technical paints would continue to be required for the future, so replacing solvents produced by the normal petrochemical route with bio-based solvents would seem to be a no-brainer way to help save the planet. It turns out not to be that simple. A much-used solvent, butyl acetate, can be produced efficiently within integrated petrochemical processes. It can also be produced from crops by fermentation to create butanol and (separately) acetic acid. These bio-fermenters look, to an outsider, like a large chemical factory – and indeed they are. They consume a lot of energy and produce waste just like any other process. When the LCA was performed as part of the project, the results were interesting. The bio-based process was better for, say, CO_2 emissions but worse in terms of water consumption and the impacts of the agricultural processes, including the transfer of potential foodstuffs from humans to chemicals. It is not at all obvious that the bio-based butyl acetate is better for the planet.

One of the frustrations of those in industry is that it is automatically assumed that "industry is bad" and that "consumers are good". A lot of the time it is the other way around. If consumers insist on buying low-cost, inefficient paints, in over-sized cans that are only partially used, applying the paint with cheap brushes onto ill-prepared surfaces and, necessarily, repainting after a few short years, then they have unnecessarily consumed large amounts of precious resource. A high quality paint plus primer combination, bought in the right quantities, skilfully applied with the correct brushes, will apparently cost more, yet will last much longer and consume fewer resources. However much the industry experts use the best-possible science to create paints that are better for the planet, if consumers insist on buying the inferior paints and applying them badly, the planet is worse off.

CHAPTER 9

Sticking in 3D

This short chapter is about the adhesion aspects of a technology that is starting to change the world – 3D printing.

The concept of 3D printing (also known as additive manufacturing) has been around for a long time. In the 1980s the industrial research group I was in gave a lot of thought to whether we should get involved. We decided (rightly) that a usable technology was too far into the future and worked on other priorities. In the early 2000s I gained access to an amazingly expensive 3D printer at the University of Leeds and was allowed to feed it a 3D file of a Celtic knot that I had created from a program I had written to generate these amazing designs. The machine used selective laser sintering of a powder, discussed below, and I remember looking at a distinctly unimpressive cube of grey nylon powder standing there when the process had finished. We took it over to a booth and started to rub away the loose powder then blasted it with compressed air. It was an awesome experience to see the knot emerge from the powder, just as Michelangelo implied about sculptures pre-existing in the marble from which they were carved (Figure 9.1).

3D printing has gone through a number of hype phases and is finally settling down to being a normal business with spurts of

Sticking Together: The Science of Adhesion
By Steven Abbott
© Steven Abbott 2020
Published by the Royal Society of Chemistry, www.rsc.org

Figure 9.1 The knot on the left was 3D printed via Selective Laser Sintering (SLS) in grey nylon. The one on the right was Fused Filament Fabricated (FFF) in black polylactic acid.

growth, ups and downs, successes and failures. It is *not* any time soon going to transform mass production, because it is too slow, the materials are too limited and the costs too high. Instead it is just another thing that lots of people use in their regular lives. They are popular for home use to make stuff, to be artistic or to encourage children to get into 3D modelling. In my formulation world, 3D printers are casually part of the lab infrastructure. In the old days if you needed a specific gizmo to attach to, say, a lab robot, you would have to find the closest approximation and spend time and money getting hold of it. Now you design the part in the afternoon, set the printer going as you leave for the night, and find the gizmo ready to use in the morning. In businesses where complex parts have to be developed and tested, it is almost unimaginable to *not* have a 3D system for generating prototypes. And in some special industries, 3D printing is now *the* production method. One specific example is dental prostheses. It would be considered strange *not* to be making these via 3D printing because each item is unique to each customer's mouth.

9.1   UV CURING SYSTEMS

In terms of this book, there is little to say about this way of creating 3D objects. Shining the right UV light at the right place and at the right time converts a liquid monomer into a solid polymer. There are various ways of doing this, none of which seem to have any relevance to adhesion. However, a few themes that appear throughout this book are relevant here:

- There is a trade-off between 2-, 3-, 4-functional acrylates and speed of cure, toughness and so on.
- There are also trade-offs around how much the system shrinks on curing – high-shrinkage is hard to accommodate in the original design, while low-shrinkage systems are generally less strong because they contain a lot of non-acrylate material.
- A system filled with particles can be both strong and low-shrinkage, provided, as mentioned previously, the (small, nano-sized) particles are nicely dispersed and can also become integrated into the cured structure via reactive groups.
- A non-robust structure can be created rather quickly, so the machine is not tied up too long with one part. Then it requires some time/temperature to allow the reactions to proceed to completion.

As with each of the 3D processes described in this chapter, the UV technique has many practical limitations. Its appeal is in allowing one-off prototypes and small-scale production of specialist parts. An example I came across illustrates this. I was visiting a small start-up company to help solve some formulation problems in a complex extrusion process that required them to pump three concentric layers of fluids through a nozzle. The shape and size of the nozzle had a big impact on the process. How could they possibly design and manufacture an optimum shape? After a day's experiments they could see what needed to be changed, the changes could be made in their computer model and then, before they went home for the night, they sent the file to their UV printer. The next day (after a heat treatment to fully cure the part), they were ready to test the new nozzle design. The important lesson for me was how entirely normal this was for this generation of developers.

An alternative UV system uses a UV inkjet printer to place drops of relevant ink in the right place, cure them instantly (though maybe not completely), and build up the multiple layers of whatever colours or materials are desired, including dissolvable materials that can be removed to allow open structures.

As with the bulk UV system, a post bake/cure cycle can improve the overall mechanical properties. The chief limitation is that although UV-curable materials are relatively versatile, the constraints on UV materials that are also "inkjetable" are quite severe, making it more difficult to obtain all the desired properties in the final object. The chief strength is that unlike most other 3D printing systems, because you have multiple inkjet heads in the system, they can contain very different materials. Therefore you can have multiple properties, for example a mixture of rigid and rubbery polymers, within the same printed part.

A very specialized form of UV printing uses a high-powered laser which does nothing to the acrylate system except at the very sharpest point of focus where a curious phenomenon ("two-photon optics") allows the system to cure. By moving the laser spot and the focal point throughout the liquid, any structure can be written. The technique is wonderful for creating sub-micron structures. After considerable effort the technique has been extended to mm-sized structures. What is amazing (apart from the sophistication of the technology) is that this ability to create exquisitely accurate micro-scaled structures has many specialist uses, especially in the bio-medical field.

9.2 EXTRUSION SYSTEMS

These systems are called FDM (Fused Deposition Modelling) or FFF (Fused Filament Fabrication). A thin strand of a meltable (thermoplastic) polymer is fed into a heated nozzle. The nozzle moves in the X–Y plane and the feed system switches from off to on whenever some polymer is required in this portion of the design. The hot liquid polymer is squeezed onto the part below it and solidifies. Once all relevant portions of this X–Y plane have received their dose of polymer, the part steps down in the Z plane and the next layer is extruded. In a two-nozzle system, a soluble polymer such as PVOH can be printed to allow the main polymer to be supported, temporarily, to create final structures that are opened up by dissolving away the temporary polymer with water. That's how the black Celtic knot in the image was made; upper loops were deposited onto sacrificial PVOH substructures.

The trade-offs are obvious and brutal. Thick strands allow a model to be built quickly, with ugly steps at the edges and unfortunate gaps within the structure. Thin strands are much slower to print, but provide smoother edges and better structures. With a small gap between the nozzle and the layer below, the polymer has a chance to make better initial contact (with better adhesion), at the cost of producing a wider strand with reduced spatial resolution. High temperatures provide faster flow, and stronger layer-to-layer adhesion because there is more chance of entanglement across the interfaces. As temperatures increase, so do the risks of excess flow, loss of resolution and distortion, plus oxidation of the polymer surface which decreases adhesion. Depending on the polymer and the design being built, the part might need to be held at a high temperature to allow maximum time for entanglement across the interface between strands. Alternatively, it might have to be actively cooled, otherwise the "solid", warm polymer sags under its own weight.

For a given polymer, lower molecular weights lead to lower viscosities that make extrusion easier, with the downside of lower entanglement across the interface, and a weaker overall object. The sorts of polymers that have high entanglement and high toughness will be generally harder to extrude than low entanglement, weaker polymers. The practical outcome of these competing factors is that polylactic acid (PLA) is commonly used, as it is a good-enough polyester, while the much more desirable crystalline polyethylene terephthalate (PET) is almost unusable except on high-performance machines.

The details of the polymer are important. Polymers have melting points and softening points. If the melting points are too high, then it isn't (usually) practical to extrude the polymer. If the softening point is too low, then the structure will tend to sag under the heat of the process. There is a third factor which depends on the polymer and its molecular weight. The higher the molecular weight, the more self-tangled the polymer is and the harder it is for it to sort itself out into a proper solid. An annoying problem for everyone is that the "same" polymer from different suppliers might have small differences in, say, branching. This means that they will solidify and entangle at different rates, meaning that machine settings that were perfect for one batch of that polymer will be not quite right for the next batch.

We even have an adhesion problem at the start. If the initial layers of the part are not well-adhered to the base plate, the drag of the filament as it is squeezed onto the layer below can mean that the part moves during printing. This might lead to a small deviation from the original, a distorted corner or, as I've seen happen, a spider's web of material filling the machine because the part became fully unstuck from the plate and the nozzle simply squirted into free space. A layer of adhesive tape can be used or, preferably, the base plate can be heated. Getting the balance of temperature to give adequate adhesion without causing slump of the part is yet another practical problem. At the same time, if the adhesion to the base plate is too good, it is hard to remove it. Although the adhesion will be modest, it is difficult to get the edge of the printed, solid part to peel, so removal is closer to being via a butt pull, which means that the effective adhesion is much higher.

All this means that filament extrusion printing of high-quality parts requires juggling of many parameters, some of which come down to entanglement and adhesion, for better or for worse. The academic literature shows that, for a typical polymer, the rapid cooling of the melted polymer onto the strand below makes it borderline whether there are sufficient tangles to give a strong bond. Practical experiments show that the strength of these parts is generally lower than desired. Specifically, although the part might appear strong when given a standard tensile (pull) test, the fracture toughness (resistance to a known crack) is lower, which is exactly what is expected when there are regular layers of zones liable to fracture because of lack of entanglement.

I must not over-emphasize the downsides. One core ability of these systems is that just about any organization can have one of these printers to produce those odd, custom widgets that can make life much easier. One lab I know uses their printer to create tiny disposable "paint brushes" used on their expensive robot and discarded after a single coat of "paint". Another creates racks to hold test tubes in just the right way to make it possible to make reproducible photos of the contents of the tubes. Neither example sounds impressive until you try to find a different way of doing the job.

At a grander scale, if you discover that each of your $million production robots needs a mechanical modification to, say, reduce vibration in operation, you can rapidly build some prototype parts to add to the robot and, when the optimal design has been achieved, print five copies for the five robots in your plant and send the 3D file to your three other plants around the world so they can print their own parts. The parts may not look as nice as the rest of the (expensive) robot, but who cares as long as the robot is working much better with an upgrade that was a fraction of the cost of getting the robot's manufacturer to make a full-spec custom part.

And maybe we'll be walking around with bones replaced by FFF structures. If you look at the range of properties of polymers compared to bones, only one commercial, safe-to-use polymer is a good match and therefore potentially usable as a drop-in re-placement. This polymer is called PEEK. It requires super-high temperatures for extrusion and is not for the faint-hearted, but with such an important potential application, research groups have been able to produce high-quality bone replacements for early testing of the technology.

At the extreme end of FFF are 3D printed homes. The extrusion is now of concrete but the trade-offs of speed, viscosity, smoothness of the walls, intermediate supports, slump, inter-layer adhesion are the same. The interlayer adhesion becomes a problem if the extrusion for some reason has to stop. As we saw in Chapter 7, concrete gets its strength from the chemical re-actions that interlink the cement and the filler materials. If a layer of concrete cures before the next layer is poured, the interface is weak because the reactions across the interface are limited.

Because the principles of adhesion are universal, it doesn't matter whether the failure is in polymer printing onto a lower layer of polymer that has cooled too much to be able to inter-mingle or in house printing from a hardened layer of concrete. In each case, a lack of entanglement across the interface means a bothersome weakness. In a 3D plastic part this might not be a problem. In a 3D printed home this might be a catastrophe waiting to happen when, say, an earthquake strikes.

9.3   SINTERING SYSTEMS

Sintering provides the excuse for writing this chapter in the first place because the process allows us to provide more detail about this fascinating method of sticking things together.

The UV method transforms a liquid to a solid. The extrusion method takes a solid through its melting point and back again. Sintering is a very different process because it joins solids together without them being liquid. Some of the classic sintered materials discussed in earlier chapters are the ceramics in our day-to-day lives such as bricks, crockery and china and the turbine blades inside some modern aircraft engines. In all cases, a powder is created, with various binders so that it is able to stick together, as a "green" form in its final shape – though not necessarily its final size because of shrinkage. The green form is fired by heating to a high temperature that first drives off any binder then causes the powder particles to come together.

The sintering process involves atoms/molecules at the contact point slowly moving by the process of "surface diffusion" to reduce the high curvature of the contact point to a bridged, lower curvature structure (Figure 9.2). High curvature is equivalent to high surface energy, so the sintering process reduces the total surface energy first by producing lower curvature and finally, with complete sintering, removing most of the surface as all the particles are joined together.

The good thing about sintering is that the initial movement of higher-energy atoms and molecules is fast, with things starting to stick together quite quickly. The bad thing is that the rate at which this happens falls off sharply because the bridges between particles reduce the curvature. It therefore gets increasingly difficult to sinter fully. Because the rate depends on the curvature of the particles, smaller particles sinter faster. They also require lots more sintering because they have so much surface.

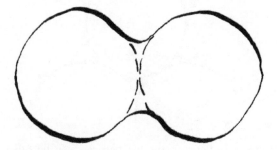

Figure 9.2 During sintering, atoms or molecules move to change the highly curved separate surfaces into a single surface with less curvature.

A good compromise, therefore, is to have a mix of sizes, where smaller particles fill the gaps between larger particles, giving the maximum packing fraction of particles and lots of opportunities for particles to form and grow bridges (Figure 9.3).

Anything that encourages the particles to move can help, though too much of a good thing would be a bad thing. An additive that was simply liquid at sintering temperatures would fill the voids but would be a source of weakness at high working temperatures as well as leaving crack-prone sharp boundaries within the main material. A good sintering aid is a flux which can melt and flow, while also dissolving or reacting with the material so that the sintering bridges are of relatively high thermal stability. These fused flux-particle bridges ideally also provide gradients of material properties that can distribute crack stresses evenly.

The Celtic knot that I produced at the university was created via SLS, Selective Laser Sintering. A thin layer of powder was spread onto a flat bed that was held at a temperature just below the melting point of the polymer. A high-power CO_2 laser beam then heated any area that needed to be fused to a solid. The energy and time were not enough to properly melt the polymer. Instead, it was enough to sinter the particles. Once one layer had been sintered, the part was lowered, a fresh layer of particles was

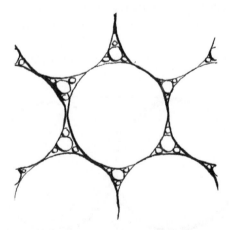

Figure 9.3   The spaces between the big particles can be filled with smaller ones, and the remaining spaces filled by even smaller ones. Less empty space means a better-quality end product.

spread, and the step repeated. Unlike the FFF knot which required a support layer of PVOH which I removed via a few hours of soaking, the un-sintered powder supports the sintered portions and the unused powder could be used for a fresh part.

The great advantage of laser (or e-beam) sintering, in principle, is that it can be used on anything that sinters – polymers, metals, ceramics. Although 3D printing is not (yet) suitable for mass manufacture, it is very much a part of high-tech manufacturing processes. It makes it possible, for example, to create turbine blade designs with complex internal cooling channels that are impossible via conventional routes. A relatively slow process is no problem when you only need a few thousand parts at the core of a business worth $billions.

There are trade-offs in sintering – a slower laser scan will give stronger sintering, but the part will take longer to create. With care and thought, the trade-offs are not so bad. If the object is sufficiently sintered to be handled and cleaned up (excess powder removed) then a conventional long, slow sintering process on the whole part can be used to generate the final mechanical strength and uniformity. The obvious objection to this is that the shape will change during sintering. Fortunately, this has always been the case for any form of sintering and it is possible to plan the initial shape to be slightly larger than the final form so the final sintering process creates the exact required dimensions.

The part that I produced many years ago was made in polyamide, nylon. General-purpose polymer parts made today are usually in the same material not because everyone wants parts in nylon, but because in practice few other polymers are printable by this technique. The problem is that you need a polymer which stays completely un-sintered at a few degrees below its sintering point and also (so you can recycle un-sintered polymer) unaffected by the high temperatures. You also need a polymer that sinters rapidly then recovers its strength quickly. The only polymer that achieves this routinely is a highly optimized nylon. It also has to handle nicely as a powder, not agglomerating as a layer is spread. To stop particles sticking together prematurely is easy – just coat them with, say, silica powder. The problem, of course, is that the silica might stop the laser sintering too. Achieving this whole package of desirable properties has been a challenge for the general advancement of selective laser sintering. Just about anything (e.g. metal parts) can be made via special equipment suitably optimized, but you must have hyper-specialized needs to justify the more sophisticated equipment.

An ingenious way around these limitations uses a humble inkjet printer.

9.4   HP JET FUSION AND METAL JET

A fascinating variant of sintering introduces another element into the sintering story. An HP inkjet printer does not sound like a promising component of a 3D printer. The ingenious idea behind the Jet Fusion process is that it takes a standard step-by-step sintering process, transforming polymer powders to a solid, but does away with lasers. Instead, the inkjet head deposits drops of plasticizer in the areas where fusion is required. A plasticizer is a molecule that enters the structure of a polymer and opens it up, allowing more polymer chain motion. Many common polymers contain plasticizers to make it easier to process them or to reduce their tendency to be brittle.

We have already discussed fluxes which speed up the diffusion at the surface to aid sintering. In this case, the plasticizer acts like a flux by making the polymer surface more mobile. When a heater plate (not a laser beam!) is passed over the layer of polymer beads, those beads without plasticizer remain un-sintered and those with the plasticizer sinter together nicely. Unfortunately, the plasticizer can flow slightly from where it is deposited. This does not matter in the centre of the area to be sintered. At the edges it would lead to partially sintered portions that would be ugly and imprecise. So drops of anti-plasticizer are deposited at the edges, a task easy to arrange with the multiple inkjet heads common to all printers. The anti-plasticizer creates clean edges (Figure 9.4).

As with laser sintering, the unused polymer particles are removed by shaking/blasting and recycled for further use. The parts themselves are too weak and need to be further heated/sintered. This increases the strength and also allows the plasticizer to diffuse through the bulk of the polymer, allowing the overall properties to be uniform, which in turn reduces local stresses and long-term problems.

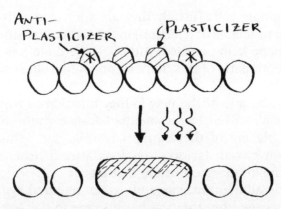

Figure 9.4 Drops of plasticizer from an inkjet printer make the polymer beads sufficiently soft to be melted by an infra-red heater, with the unplasticized beads unaffected. Drops of anti-plasticizer stop migration at the edges of the plasticized area.

The Metal Jet version creates a "green" ceramic through the simple idea of ink jetting a solution of an adhesive polymer ("binding agent") onto consecutive layers of metal powder. Heat allows the solvent to evaporate and enables the polymer to properly bind to the metal powder. Once the part has been printed, the excess powder is removed and recycled, revealing a green part with sufficient strength to be handled gently. Then the green part is sintered just like any other metal (or ceramic) to produce a full-strength part.

9.5 RE-PRINTING HUMANS

If you need a replacement kidney you currently have to hope that a matched one becomes available in good time. In the future you might expect to have one printed for you.

There are three major challenges before this becomes routine. The first two are, relatively speaking, easy, even though they sound hard. The third, which sounds easy, is really hard:

1. Build the general shape with some suitable temporary material such as a fibrin (the blood-clotting protein) or an alginate (a food polymer that can be set to a solid with calcium ions), using some sort of inkjet or extrusion or electrospinning technique. At the same time, place the

appropriate cells (ideally they are your own cells so there are no problems of rejection) in the right 3D position within the shape being created. Although there are lots of issues with both parts of this, making organ-like objects is readily do-able.

2. Make all parts of the new kidney function as they should. This isn't as hard as it sounds because given an approximate placing of the appropriate cells and infusion with relevant growth factors, they look after themselves (that's how the kidney grew in the first place).

3. Provide all the nutrients and oxygen required by the organ as it moves from the first basic shape to the second functioning organ. This is genuinely hard because it requires a full system of vein-like tubes running through the prototype organ, delivering a regular stream of the needed nutrients. If this could have been done by letting the organ sit in a vat of fluid, we would have advanced much further than we currently have. But this passive diffusion works only on the very small scale and cannot produce something like a kidney.

The overall adhesion challenge is to get the cells to stick together well-enough to keep the physical structure intact while allowing the shape to change during the natural progression of cell death (wrong cell in the wrong place) and cell growth to take place as the organ develops into the fully-working version. In the final chapter we find that there is a big debate about whether cell-to-cell adhesion takes place via special adhesion molecules with special adhesive interactions, or whether adhesion is just standard van der Waals interactions taking place at positions controlled by those adhesion molecules.

9.6   THE FUTURE FOR 3D PRINTING

The majority of 3D-printed parts are made from a single material, often with only modest performance. The range of printable materials is surprisingly small, often because of the constraints imposed by the reality of creating strong interfaces via entanglement of extruded or sintered materials. The majority of real-world parts are made from multiple materials, each optimized to the required performance. This means that 3D printing is very far from being a general-purpose manufacturing technique, even if the relatively low speed and relatively low surface quality could each be improved.

On the other hand, 3D printing can produce parts that are impossible (such as my knots) or impractical to obtain (such as special test-tube racks or experimental extrusion nozzles) via conventional routes. It is these new capabilities which ensure that 3D printing will continue to grow in volume and capabilities.

Whatever its future, those developing future generations of machines will have to come to terms with the basic laws of entanglement and dissipation of crack energy. By definition, "additive" manufacture involves sticking things together. A sound knowledge of what helps and hinders the creation of strong interfaces will always be required.

CHAPTER 10

Not Sticking

In the list of problems facing the world, the problem of how to get ketchup to flow out of its bottle is not especially serious, yet when anyone comes up with a potential solution to the problem, there is great media excitement. As far as I know, most of the ideas have been practical no-hopers. Universities have whole departments dedicated to taking the results of an experiment and blowing them up to "saving the planet" announcements, often (though sadly not always) to the consternation of the researchers themselves, who know how unrealistically exaggerated the reports are. These non-solutions have their five minutes of fame and are then quietly forgotten. Recently, however, encouragingly impressive YouTube demos have appeared of solutions to the ketchup problem which are both smart and scientifically sound. They do appear promising, but whether they will make it into production depends, as with all products, not on the glamorous 80% of the solution (the smart idea and the enthusiastic video) but the dull 20% (cost, safety, aesthetics, testing all possible usage scenarios, patent wars...) which always takes 80% of the effort.

These new solutions address a fundamental law of physics which prevents our ketchup from pouring out from the bottle as cleanly as we would like it to. To start with, we can imagine a bowl full of ketchup. As we tip the bowl, the ketchup starts to flow.

Sticking Together: The Science of Adhesion
By Steven Abbott
© Steven Abbott 2020
Published by the Royal Society of Chemistry, www.rsc.org

What we can observe is that the uppermost surface of the ketchup moves faster than the layer nearest the surface of the bowl. If we could look close to the surface we would find that the flow has slowed down tremendously. Returning to our bottle of ketchup, the free flow in the centre is fast, the flow at the walls is slow (Figure 10.1).

The law of physics at the heart of our problem is the "no-slip condition," which says that the velocity of the liquid at the wall surface is exactly 0, leaving us with a stagnant layer that cannot be persuaded to move. In something like ketchup, which has its own entanglement to provide a desirable semi-solidity, the no-slip layer holds up the rest of the flow.

The clever solution is to replace the wall of the bottle with a liquid. There isn't a no-slip condition between liquids so the ketchup can flow freely. Now of course, liquid bottles are going to be a hard sell: the real trick is to design an LIS – Liquid-Impregnated Surface – a micro- or even nano-rough surface with an integrated, not-too-viscous, liquid which hardly evaporates, which fully fills/wets the surface and doesn't mix with the ketchup. Any mixing of the liquid into the ketchup would raise issues of food safety and flavour, and any mixing of the ketchup into the liquid would tend to reduce its efficacy (otherwise the liquid could just be ketchup). At the time of writing, the patents from one such company, LiquiGlide, propose that the surface of the bottle is sprayed with food-safe carnauba wax, which forms a micro-rough surface. The surface is then coated with food-safe

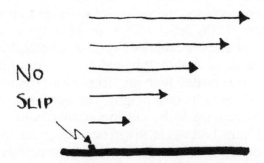

Figure 10.1 Liquid away from a surface can flow quickly. The velocity decreases close to the surface and becomes zero at the surface. This "no-slip condition" has many implications.

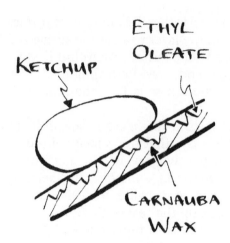

Figure 10.2   Ketchup sticks to the walls of a normal bottle. Here it floats on a
thin layer of an oil (ethyl oleate) which is held in the nano-rough
surface of carnauba wax.

ethyl oleate (an oil). The ketchup then glides easily along the
liquid surface (Figure 10.2).

The no-slip condition also has a profound effect when we try to
clean a surface. What we want is for our cleaning fluid to get
right to the surface and remove the soil, yet this is impossible
because the fluid cannot flow there. The reason we use micro-
fibre cloths or add particles to our cleaning fluids is that the
microfibres and particles can break through the no-slip boundary and force the soil to come away.

There is a more serious problem to be solved for those who live
in, say, Minnesota where the winters are cold and snowy. The
problem is not so much the weight of snow on your roof – snow
can be relatively easily removed. The problem is ice. If this builds
up on a roof (or electric cable or any other fragile object) then its
weight can be catastrophic. It would be wonderful if all critical
surfaces could be coated with an "icephobic" treatment so that
the ice slid off under its own weight. The no-slip condition is the
reason the ice cannot slide; putting a smart liquid coating, as
with the ketchup bottles, is theoretically a great solution. The
idea is hopeless in practice because all those surfaces that need
treatment could not be treated with, say, carnauba wax and ethyl
oleate; nor, in any case, would the combination survive outdoor
exposure.

An elegant alternative solution was inspired by a paper describing a set of beautiful experiments that were aimed at solving one of the key mysteries of PSAs: why do silicone release layers work so well? When I first read that paper on silicone release, I was thrilled that some bright scientists could do something so elegant. When I later came across a paper on a novel way of solving the ice build-up problem, I was thrilled that the authors had been inspired by the same papers that I had read, and had done something imaginative with the same ideas. That is science at its best. We start with the silicones then return to making icephobic surfaces.

10.1 SILICONE MAGIC

If you read just about any other paper on silicone release liners/ papers you will find that they "explain" the release effect by referring to silicone's low surface energy. This is despite the fact, known for decades, that other surfaces of the same low surface energy provide no such light release. Try to stick some adhesive tape to your Teflon frying pan (a fluorosurface). With my tape, instead of the 250 g pull-off resistance on stainless steel I measured 90 g on my very high quality and very non-stick frying pan. So, yes, a PSA doesn't stick well to Teflon but even that 90 g is a massively higher pull-off force than measured on a silicone release liner where I found it difficult to measure even a few g with the same tape. With a stronger tape I could lift the whole pan, again not something I could have done if the pan had a silicone release surface.

The key to silicone release was found in the amazing experiment that cracked the problem. To make sense of it you have to recall the diagram (Figure 10.3) which showed the PSA being peeled apart:

About 1 mm ahead of the peeling zone there is a compression zone. Although it is not obvious, the fact that this zone exists means that the adhesive is pushed to the right as the compression arrives, then moves back to the left before the peel zone arrives.

In the key experiment, some fluorescent beads were added to the adhesive. The beads in the middle of the adhesive layer followed the right-left motion as expected. On a Teflon-type surface (or, indeed, any other normal surface) beads in contact with the surface showed zero motion – after all, this is the no-slip

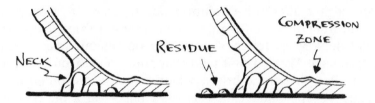

Figure 10.3 When we previously looked at this image, the focus was on the fingers. Now we switch to the strange fact that the stresses move ahead of the peel front compressing a region ~1 mm ahead. This is important for understanding how silicone release films really work.

zone (Figure 10.4). The fact that there was differential motion (the inner bead moved, the surface bead didn't) means that some of the crack energy had gone into sliding the adhesive past itself, dissipating the energy; anything which dissipates crack energy helps to reduce the chances of the crack separating the surfaces.

On the silicone, the behaviour is different. The particle at the interface moves, slips, just like the particles in the middle. This means that there is no differential motion and no dissipation of energy, which in turn means that the crack can happily rip things apart.

The silicones have no adhesion because, uniquely among pure polymers, the whole polymer moves past itself with no problem.

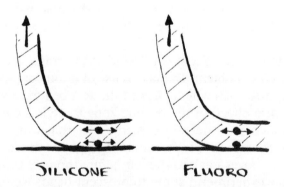

Figure 10.4 The surface of a silicone is liquid-like and all parts of the adhesive can oscillate together as the compression zone (not shown) passes through. On the fluoro- (and any other surface), the adhesive cannot slide at the interface so crack energy is dissipated by relative motion within the adhesive.

It is equivalent to having a layer of liquid at the surface of the polymer, so the no-slip condition doesn't apply in this case. This is why silicone release liners are used extensively and fluoro-based release liners do not exist. The reason that non-stick frying pans use fluoropolymers rather than silicones is discussed below.

Why are silicone surfaces pseudo-liquids? Because the siloxane bond along the polymer rotates with exceptional ease. Conventional polymer chains cannot move this easily. With polymer chains free to rotate and wiggle, the silicone surface is closer to a liquid than a solid.

Using this insight that the silicone release surface is a semi-liquid, the icephobic team decided to test a range of surfaces that could act similarly, by adding liquid "plasticizers" to the solid polymer. It was easy for them to create rather weak semi-polymer surfaces that showed wonderful icephobic properties, though they were too liquid to survive real life. They could make tougher surfaces that could survive real life, without being sufficiently icephobic. They could also create plenty of surfaces with a credible balance of easy release and practical toughness, using a wide range of polymer/plasticizer combinations that could deliver the required slipperiness. In their experiments, the ice really did fall off!

I doubt that the icephobic work will lead to ice-free roofs in Minnesota, but I don't hold that against the authors. The basic idea has a lot of potential in other areas, including novel PSA release tapes. Many scientific papers include an eye-catching motif around a tough general problem to draw people into the challenge of reading and learning something new. The good ones, such as the icephobic paper, bring out the good science that is relevant to that general problem, and indicate future possibilities without over-claiming. The bad ones tend to over-claim *and* include rather unexciting science that doesn't lead anywhere.

10.2 OTHER WAYS TO REDUCE ADHESION

Through knowing how to make adhesion work, we can now say how to make adhesion *not* work. For example, we can reduce adhesion by:

- Making the surface too rough for the adhesive to cover it

- Making the surface impermeable and free of interactive functionality
- Making sure that the surface is contaminated with oil, dirt, particles or a weak oxide layer
- Making the surface too cold or too hot
- Making the adhesive flow too easily – or making it too solid
- Applying forces to the interface very, very slowly or very, very fast
- Removing or destroying any surface functionality that could react with the adhesive
- Adding too many groups or too much adhesion promoter and thereby create a brittle bond
- Loading the joint in a manner for which it is not optimal – in peel for a shear joint, at an angle for a butt joint
- Adding defects at critical points in the joint – usually at the edges where stresses are concentrated

What is important about that list is that most items aren't to do with the adhesive itself. This is a reminder to always think of the whole system if the issue is avoiding or efficiently achieving poor adhesion.

10.3   LUBRICATION

There is one prime rule of lubrication: don't let the two surfaces touch. If the surfaces aren't touching, there's no friction between them. The remaining friction is the viscous drag of the fluid between the surfaces, the lubricant.

The lubricant with the lowest viscosity is air. Those old enough to have used a computer hard drive will have had a super-smooth read/write head flying (literally) over a super-smooth magnetic disk. The aerodynamics of the head were crucial to ensuring a minimum height (for highest density recording) with near-zero risk of crashing. Because near-zero wasn't quite the same as zero, a tiny layer of lubricant existed on the surface of the disk. All this was possible because the entire environment was controlled, and the disk rotated at a constant speed.

The interfaces we encounter in normal rotating or sliding machinery are far less controlled, which is why air lubrication is not an option. Now we have a big set of compromises to make.

A viscous lubricant, for example, will resist being squeezed out of the moving assembly, but will provide a larger drag. The viscosity of a perfect lubricant for running cold will be too low when the assembly becomes hot. In any case, there are likely to be shock loads that will temporarily squeeze out the oil, leading to contact and wear. It is possible to reduce that effect by mixing in a backup lubricant, which is usually something tough and stable like molybdenum disulphide, a mineral that comes in the form of sheets or flakes that slip just like graphite.

Another trick at an engineer's disposal is to ensure that the surfaces have a controlled "nano-roughness", just enough to contain a sheet of oil. As in the case of the ketchup bottle, this helps to maintain a slippery boundary layer.

10.4   NON-STICK HIP JOINTS

Given that any smooth or compliant surfaces that get into contact can stick reasonably well, and given that our bodies contain plenty of those, the question arises of why we aren't one solid mass. That is easy to answer – as long as various independent parts or organs can arrange adequate surface roughness, there will be no adhesion. But in some special cases, we want highly smooth, highly mobile interfaces.

Of all the moving joints that we can imagine in engineering – from automotive suspension arms to robotic manipulators, the human hip joint is one of the most challenging. As well as having to withstand the complex forces and high stresses of walking, it is also where we need super-smooth surfaces that don't stick to each other. Millions of years of development resulted in an additional, beautifully simple trick used by biology that can reduce adhesion to an immeasurably small amount – it is so simple that to understand it we just need two pieces of smooth rubber, some water and some dishwashing liquid. A smattering of fundamental physics rounds off the experiment.

As we know from Chapter 4, if we put two strips of rubber together, we can measure a peel strength of, say, 40 mN m^{-1}. If we repeat the experiment with some distilled water on the surface, the measured value drops to around 4 mN m^{-1}. Careful experiments with high-tech equipment reveal why. The water molecules refuse to get completely out of the way and their thickness is enough to greatly reduce the van der Waals attractions between the two rubber surfaces. Now repeat the experiment with a tiny amount of surfactant – our dishwashing liquid – added to the water. The peel force is not just reduced to zero, it has become negative, i.e. the surfaces repel each other.

The repulsive physics that create this scenario depends on the fact that surfactants have long hydrophobic tails that like to sit on the surface of the rubber, and charged (ionic) heads sticking out into the water. As the two surfaces come together, the charges come closer and, because like charges repel, the repulsion increases.

Instead of dishwashing liquid we have synovial fluid within our joints and this uses surfactant-like polymers in exactly the same way. The molecules sit on the surface of the joint with charges sticking out into the fluid, repelling the other surface and preventing our hips from sticking.

The theory behind this is called DLVO after Derjaguin and Landau who came up with the idea in Russia and Verwey and Overbeek who came up with the idea in Belgium, unaware of Derjaguin and Landau because scientific exchange was complicated by language issues and the Second World War. The same subtle techniques that can measure the effects of water molecules between surfaces can measure the surfactant effects which exactly match DLVO theory up to the point where the individual surfactant molecules start getting in the way.

Another means by which nature reduces bio-adhesion and friction is somewhat similar to the Liquid-Impregnated Surface system that allows ketchup to flow freely from the bottle. I first came across it in a paper titled "From Red Cells to Snowboarding: A New Concept for a Train Track" which included a design for a train track based on goose feathers. Although the authors could have chosen to write a conventional paper, by adding a strikingly silly idea, backed up by serious science, their interesting ideas stood a much better chance of being understood and passed on to others – just like the icephobic paper. The Red Cells part of the title is there because a classical analysis of red blood cells inside our capillaries demonstrates that as the cells are approximately the same diameter as the inside of the capillary they will get stuck by friction. The trick adopted by nature to stop the cells from sticking is the same as a snowboarder's. If you stand on a snowboard in soft snow, you sink into the snow because your weight pushes all the air out of the snow. If you slide downhill, the board still tends to sink in, but the air being displaced can't get out of the way fast enough and forms a cushion under the board. The optimal snowboarding experience is not one of sliding on snow but of sliding on air.

The red blood cell equivalent is that the walls of the capillary are made from a loose polymer network filled with water. As the cell squashes the system, the expelled water is forced into contact with the cell which, therefore, floats through the capillary on a cushion of water (Figure 10.5). If you do the calculations with a heavy train moving along a train track made from goose down, you find that it could slide along with a glorious absence of friction.

Once you have finished your snowboarding run and come to an elegant stop in a plume of snow, the board will once again expel the air and sink into the snow; you can pick up the board and walk away. For the red cell, the sliding continues until it reaches a large vein where it can float freely. For the goose feather trains, the problem of what happens if they stop requires, as the authors admit, "further research".

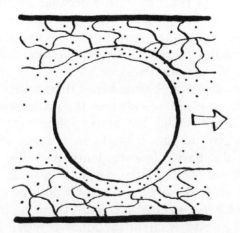

Figure 10.5 A red blood cell floats through a capillary on a layer of water (small dots) squeezed out of the loose polymer network attached to the wall of the capillary.

10.5   NON-STICK FRYING PANS

To stop stuff from sticking you don't, as the myths suggest, need a low surface energy surface – the practical adhesion through mere van der Waals attraction varies only by a factor of two between a "low-" and "high-" energy surface, not enough to make a big difference on a frying pan.

Instead, what is needed is a surface which is highly crystalline and into which no molecules will want to intermingle. Also, there must not be any type of chemical bonding between the surface and the hot food, so the surface has to be very inert. Under that definition the only practical material which is also tough, affordable, safe etc. is polytetrafluoroethylene, PTFE, commonly known as Teflon. The chain of carbon atoms that make up the polymer backbone is covered with fluorine atoms and these fluorocarbon groups dislike everything else in the universe that isn't fluorocarbon. Any molecule that comes close is sufficiently unhappy in the fluoro environment that there is no attraction or intermingling. These fluorocarbon groups also crystallize together to create an impervious surface. The fluorocarbon bond is very strong and leaves no opportunities for molecules to react with the polymer. This means that there are zero chemical bonds across the interface. PTFE, therefore, is perfect for non-stick performance in terms of adhesion science. It is also perfect in practical terms because the stability of the polymer allows it to be used at high temperatures on the surface of a frying pan. Up to the highest practical frying temperatures, the PTFE does not react with oxygen or food and therefore it retains its chemically inert surface.

Why are some non-stick pans better than others? A cheap pan with a thin coating will quickly lose the protection via mechanical damage. If the coating has been applied badly it might have poor adhesion to the pan, it might have nano- or micro-pores, and it might also have a few accidental chemical bonds on the surface which are opportunities for food to stick.

This brings us to the old question: "If nothing sticks to PTFE, how does it stick to the metal surface?" There is no mystery here. If you can get just a few chemical bonds between the metal and the PTFE, adhesion science tells us that that is more than enough for good adhesion. Heat *anything* hot enough and it will start to react, so the trick is to apply the polymer at a temperature where the cleaned, activated, hot metal surface is able to make those few chemical bonds that are required. PTFE itself is hard to make into a coherent film because it cannot be applied from a sensible solvent and has a high melting point. Fortunately, the high temperature required to react the PTFE is enough to create, via sintering, a good, pinhole free, tough layer from a coating of PTFE particles supplied as an emulsion. When this is done properly, the coating remains non-stick for many years of cooking. Because the PTFE is sintered, a poor sintering process will leave the nano- or micro-pores mentioned above.

Why don't we use silicones, given that they are brilliant for non-adhesion? Because their highly-desirable flexibility which gets around the no-slip condition makes them way too soft for a frying pan. Can't we make them a bit tougher? Yes! As we discussed in the introduction to silicon-based chemistry, the flexible release silicones are siloxane-based while an alternative silicone system, based on the delightfully-named silsesquioxanes, is rigid. As you go from easy release pure siloxane systems to relatively hard silsesquioxane systems you get some interesting materials, but they are neither non-stick nor especially hard so they can never be used for frying pans. My favourite spatula for cooking is a "silicone" one; it has a good balance of strength and flexibility and is relatively easy to clean. However, it's certainly not "non-stick" (picking up a cake mix with a non-stick spatula would be tricky!) and it might well be a mix of siloxanes and silsesquioxanes.

What about seasoned cast iron or carbon steel pans? We all know the stories of grandmother's iron pan that stays perfectly non-stick as long as it is never cleaned with washing-up liquid and which rejuvenates its non-stick capabilities each time it is used to fry fatty foods. Heating any oil hot enough makes it decompose and react into a black carbonized substance. It is funny to read the confident assertions where one person says that fat X is good "because ..." and another says that fat Y is good "because ...". My view is that these are pseudo-reasons without much point because there are so many variables in, say, preferred temperature for the process. What matters is getting the first part of the Teflon story right which is a highly impermeable coating which does not allow anything to penetrate and intermingle. A highly carbonized surface can achieve this. The second part, a surface that cannot react with the food is clearly not the case because these messy carbonized gunks will have plenty of reactive groups. Without this second part, the pans have to gain their non-stick capabilities via a different process. A clue to how it all works is provided by a different puzzle.

The puzzle is the oft-stated prohibition on the use of dish-washing liquid to clean these pans. How can a little bit of surfactant undermine grandmother's decades of hot cooking given that surfactant can't remove the carbonized material? The washing-up liquid must be removing something else, which is most likely to be oil trapped in tiny pores in the blackened surface. Where else have we seen oil trapped in tiny pores? In those easy-glide ketchup bottles. I think that grandmother's secret is that the blackened material itself is impermeable like Teflon, but it has pores within it that hold enough oil for the food to glide on the surface. The "don't use washing-up liquid" prohibition might be somewhat of a myth because a fresh coat of oil after cleaning should be able to penetrate the pores. However, if the pores are opened up and water gets in and the pan isn't properly dried before oiling, then the patina is under attack. Better to leave the oil in place and add a little extra to flow into any part where the oil has been removed.

We know that the coating is highly permeable because if these pans aren't dried properly (some users heat them on the stove after washing), water can easily get in to rust the iron. Even the tomato acid in a cooking lasagne can be enough to attack the iron, not something that would be possible if the coating was properly impermeable. In either case, once water/acid attacks the surface the protection is destroyed and the whole process of protecting the pan has to be re-started.

I prefer to go with the science which says that impermeable and unreactive Teflon, if well-adhered to the pan, will provide years of good cooking without all those energy-hungry 250 °C for 1 hour treatments to create the pan coating. If you enjoy cooking on patina-protected pans, then take pleasure in the fact that the non-stick properties are more likely to be due to holes in the patina than to the patina itself.

10.6 UNSTICKING

A big set of questions about adhesion revolve around how to unstick things. We have all the answers in general. Our problem is knowing which of the possible solutions will do the job.

Clearly peel is always going to be more successful than shear or straight pulling.

If something is too thin to get underneath to start a peel, then a strip of strong adhesive tape might work well. As the video demo shows, if you cannot start to peel the end of a roll of tape, it can be pulled off by applying a strip of tape around the free end.

https://youtu.be/8w8TgP1Kz40

Most systems cannot resist a short, sharp shock so a super-fast peel will often do the trick. Unfortunately, if we are not quite fast enough, we usually break something else (such as the paper backing of a label). You can really impress your friends by saying, as we discussed in Chapter 4, "In physics, time is equivalent to temperature, so lowering the temperature can often help the adhesive to crack off easily, as if we had done it super-fast". If we all had access to liquid nitrogen, we would adopt this trick routinely. Dry ice would also be good. In reality, we may have to manage with something from the freezer. An ice cube, as it melts, increases the thermal contact, helping to lower the temperature quickly, and with the right sort of adhesive the water might also help to weaken it. To avoid a mess from melted ice, something metallic stored in the freezer for the purpose of removing some adhesive can work impressively well. An alternative trick is to spray the joint with one of those cans of cleaning air. The rapid expansion from the nozzle leads to rapid cooling of the joint and, as you can find in YouTube videos, easy breaking.

If the adhesive is solvent-based, then allowing solvent to soak through to the interface will help. There are two problems with this:

1. The solvent might damage the surface
2. Most homes do not have a good range of technical solvents

242

Chapter 10

The first problem is solved by finding a spot where a solvent check can be carried out without risk of visible damage if it goes wrong. The second requires some ingenuity. Strong vodka might provide enough ethanol. Nail varnish removers (often acetone or ethyl acetate) might work. Some aromatherapy oils (e.g. limonene) can be quite aggressive solvents – which is a reminder that just because something is "natural" does not mean that it is necessarily safe to use. Whatever you try, allow time for the solvent to get through to the adhesive. Trying too hard, discussed below, can be counterproductive.

The advantage of the old-fashioned collagen glues, as mentioned before, is that the best solvent for them is water, so they are readily softened with heat and moisture for easy repair of veneer or musical instruments. The disadvantage is recounted in stories of musicians who have paused for a concert rehearsal in a rather damp venue, left their instrument in a spot where the sun happens to reach the instrument and the combination of heat, humidity, and the tension of the strings, leads to an unplayable instrument at the end of the rehearsal break.

Removing paint used to be easy – use lots of dichloromethane (methylene chloride) which is an astonishingly good paint remover because it happens to have a good solubility match to many paint resins and is a small molecule that can penetrate easily. The safety profile of dichloromethane is now deemed to be unacceptable, making it necessary to find alternatives. There are many such alternatives out there, all claiming (many of them wrongly) to be safe and effective. I have been involved with a few of these replacement solvents. The safer ones are slow to act and less volatile which at first seems a problem. The trick is to turn the problem into a solution. With dichloromethane you had to work fast because although it can open up the paint, it can evaporate so quickly that the paint re-forms. This means that you have to work on small patches. With the slower, less volatile systems, the whole surface can be treated, you can go off for a coffee break, then return to remove the paint.

The obvious problem of solvents not attacking the paint has a non-obvious counterpart of a solvent that is too good and makes the paint soft and stringy, and very hard to get rid of. The best is a solvent that just weakens the paint, while keeping it relatively brittle so it is easy to scrape off. A sharp, weakened, interface is

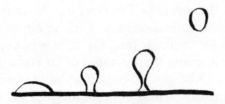

Figure 10.6 Oil on clothes or dishes can, with the right surfactant, be made to roll up into a ball and float away.

easier to remove because a crack can run along it rather than getting dissipated into a too-soft paint.

When we have a layer of oil sticking to a cooking utensil or a dish, we can lift it off with a dishwashing liquid. The usual explanation provided is that the surfactant in the liquid lowers the surface tension and that the oil is dissolved in the surfactant "micelles", clumps of surfactant molecules sitting in the water. The real explanation is much more complex (so dishwashing formulations are also complex) because there needs to be a special match between the oiliness of the oil and the behaviour of the surfactant and water. With this match, the drop of oil "rolls up" from the surface (Figure 10.6). The theory behind this is called HLD-NAC and is fully described in my free Surfactant Science eBook.

For other food residue on pans and dishes there are four factors: Chemistry, Temperature, Time and Mechanical Action (Figure 10.7). It is amusing to English speakers that this classic

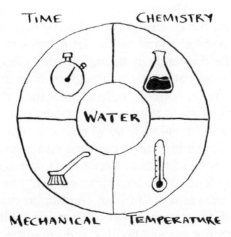

Figure 10.7 The Sinner Circle. To unstick the soil, you can choose any combination of the four factors.

description of the options for cleaning is called the Sinner Circle – with the implication that it is the sin of uncleanliness which needs to be fixed. In fact, the idea was popularized by a German scientist, going by the name of Dr Hubert Sinner.

Many of you will be glad to hear that hand washing of dishes results in an unsatisfactory combination of all four, with no single factor standing out to be able to help us, as we cannot handle strong chemicals or high temperatures, we do not have much time (other than soaking) and can only provide so much mechanical energy before our patience runs out.

If you have to choose one factor from the Sinner Circle, then you won't be surprised to find that my advice is to make it Chemistry. This is by far the most effective and is the one chosen to a significant extent in household dishwashing machines and to a full extent in commercial machines. The key element is caustic/alkali chemicals such as caustic soda. These attack the oils (breaking them down into glycerol and soap molecules), the proteins and the starches, meaning they get rid of almost everything. In household dishwashers the concentrations of these aggressive chemicals are relatively low and some use is made of surfactants. However, it is very hard to find a surfactant that is effective and which also doesn't foam under the strong jet action needed for good cleaning. Because both chemical and surfactant levels are relatively low, we need relatively long wash cycles to get the job done. Commercial dishwashers have only minutes for a wash. Therefore they use high temperatures and strong caustic for the cleaning, backed up by whatever chemicals are needed to avoid clogging up the jet nozzles by the hardness in the water. The downside of strong caustic is that it can attack some forms of glassware, causing the glass, over time, to change from clear to cloudy. For commercial machines, the trick is not to use the types of glass that are affected by caustic. For home machines there is a balance between how much caustic the manufacturers dare add to their cleaning tablets and how careful the users are to not put in glassware that is damaged by caustic.

Why do many foods stick so well to dishes? They are just like any other glue that at one time was liquid and dries out in perfect contact with the surface. They are not especially well stuck – but equally, they not especially thick and are not especially hard, which as a combination makes them frustratingly difficult to remove by scraping. Thicker deposits are easy to scrape because you can exert a strong peel force on them with the scraper. We cannot easily apply peel forces to thinner layers (finger- and thumbnails are surprisingly effective, because they are so thin and can flex to apply the forces in the right directions). If the residues are still quite soft (the classic example being oat flakes from muesli or porridge) then the crack energy is dissipated when the protein or carbohydrate molecules slide past each other. As we know from PSAs, dissipation and adhesion are enhanced via the right level of crosslinks. For cooked foods there are plenty of opportunities for chemical reactions that create enough crosslinks to make the food frustratingly well-adhered. Fortunately, when faced with a strong, hot, caustic solution such molecules rapidly fall apart; what we could not remove in warm detergent is easily removed in the dishwasher.

10.7 THE END OF LIFE FOR ADHERED SYSTEMS

If an object falls apart because of a poor adhesive, the planet has the problem of extra unnecessary waste. If an object is hard to break into its individual components at its end of life because the adhesive is too good, then the planet has the problem of mixed waste. Adhesives can't win – they are bad if they're bad and they're bad if they're good.

Why not insist that all adhesives are switchable, providing perfect bonding during their working life then easy de-bonding once they are done? Well, it's easy to insist – but how, in principle, should it be done? Promoting "de-bondability" with water means that the adhesive might fall apart in high humidity. De-bonding with UV light means failure in sunlight. De-bonding with heat means an inability to work in hot environments. De-bonding via some chemicals added specially to the adhesive means that for 99.9999% of the adhesive's life these chemicals have contributed nothing to performance so are, arguably, wasted resources.

What about insisting that all adhesives should be "natural" and "biodegradable"? Just as there are few useful biodegradable polymers, there aren't many examples of natural adhesives that have the properties we require for the super-thin, super-efficient, super-long-lasting adhesives we prefer. And biodegradability is surprisingly hard to achieve in any material with respectable adhesive qualities – none of us wants an adhesive that opens up in water and is readily consumed by bacteria. Very often, the best that can be managed with a bio-based material of moderate capabilities is "compostability", i.e. it can be destroyed over weeks in an industrial composting facility.

In any case, the problem isn't primarily one of getting rid of a few microns of adhesive; rather the problem is getting the adhesive off both surfaces. The energy required and process complexity needed to remove such thin layers of material will generally be unjustified in terms of rational green principles.

If we change the focus from the adhesives to the adherends, then we already know that otherwise pure polymers containing less than 5% of other polymers such as inks, barrier layers (for example, the EVOH in PE packaging films) or adhesives are considered by the regulatory authorities as adequate for re-cycling into the original material. For metals it is surely not a problem in general to burn off excess adhesive along with other contaminants such as paints.

Saving the planet is hard. If we focus precious resources on the issues of effectively repurposing the adherends then, in most cases, the low levels of adhesive will be an insignificant problem.

CHAPTER 11

How Nature Sticks Things

In Chapter 10 we made the detour from adhesion to non-adhesion, and encountered the hip joint, which we don't want to stick. Now we explore how nature solves problems like sticking a mussel to a rock in the ocean, a fly to a spider's web, or holding bodies and organs together.

Nature has had time to consider the matter and has come up with fascinating solutions that we investigate here.

This has inspired plenty of academic research into creating "bio-mimetic" adhesive systems, on the grounds that nature often finds greener, gentler solutions to problems than humans who tend to use "chemical" approaches. As far as I can tell, none of these efforts has produced a solution to a major problem, and it is generally the case that artificial adhesive systems have a superior balance of price, performance and even greenness than bio-based adhesives. I find bio-adhesive systems fascinating not for any special merit in helping us replace conventional adhesives, but because the science that biology has invented is often unexpected and inspirational to fresh types of thinking.

11.1 A BIT MORE ON GECKOS... AND COCKROACHES

There is a wonderful story from Prof. Kellar Autumn about months spent failing to get a simple experimental result.

Sticking Together: The Science of Adhesion
By Steven Abbott
© Steven Abbott 2020
Published by the Royal Society of Chemistry, www.rsc.org

As discussed in Chapter 3, the final part of the gecko's hierarchy of structures is the nano-scaled spatula, where all the adhesive contact takes place. The aim was to place a gecko spatula into contact with a test surface and measure the force needed to pull it off. The experiments kept finding that the force was zero, which was most unfortunate. Eventually, they thought "let's do it like a gecko and not like a scientist" and after placing the spatula onto the surface, they dragged it back slightly at an angle. That trick gave the expected strong adhesion. Why would the spatula only work at a strange angle? When they did the calculations, it all made sense. If a gecko has to spend some time hanging upside down, it would ideally expend as little energy as possible. When you work out all the angles, the best adhesion with the most friction and least energy is the angle of tilt found (accidentally) by the experiments. Adhesion is a property of the system, and in this case, the needs of the system (low expenditure of energy) dictated the specific attachment angle for maximum adhesion.

Everything in life is a trade-off. The angles of the spatulas are wonderful for climbing and hanging around. This means that they are poor for giving grip during descent. Fortunately, geckos have another trick. They can rotate their hind feet so that the angle problem is not too severe.

It turns out that cockroaches can climb using the same angle-dependent trick used by the geckos. Unfortunately, they cannot twist their feet during descent. How, then, do they manage to descend safely? Whatever the correct term is for a cockroach's knee, the trick is to use pads on their knees, that are arranged at the opposite angle to those on the feet, automatically providing the appropriate grip during descent. This trick was worked out by Prof. Walter Federle who also solved a different deep mystery via an ingenious technique.

11.2 TREE FROGS INVENT TYRE TREADS FIRST

Tree frogs can stick happily onto wet leaves and stems. Careful examination of their toe pads revealed a structure that seemed to allow a special tree frog glue to come to the surface (Figure 11.1).

Prof. Federle and his team wanted to know how sticky this glue might be. One way to determine this is to measure its viscosity. As discussed in Chapter 2, we are familiar with the fact that

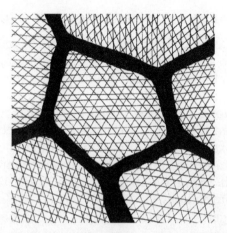

Figure 11.1 The pad of a tree frog is made of structures separated by gaps that allow the water to escape like tyre treads. The structures themselves, just like the gecko, contain sub-structures.

water flows easily and is low viscosity, honey flows less easily and is viscous and sticky, and treacle is even more viscous and sticky. It normally takes a few ml of liquid to be able to measure a viscosity. You can imagine that it is rather hard to get hold of a few ml of the mysterious glue in the tree frog's toe pads. To be able to measure the viscosity of a tiny sample of the frog's toe liquid, they used a trick called an optical tweezer. They put a tiny sphere into the liquid and shone an intense laser beam onto the sphere. The strong electric field in the laser beam is enough to trap the sphere in the centre of the beam. When the beam is moved, the sphere moves with it, though with a lag caused by the viscosity of the liquid. At the end of these elegant experiments they could announce that the viscosity of the liquid was... the same as water. It wasn't a glue at all. In fact, the story is the other way around. The pads were not providing a liquid for the surface, they were acting just like the treads on a tyre to allow water to be quickly taken away so that the pad could get into close contact with the leaf and establish, just like the gecko, good surface energy adhesion. Having discovered tyre tread technology, the frogs also worked out Stefan's law and its R^3 dependence on the size of the drop needing to be squeezed out. The large number of very small sub-pad structures means that the drops can be more easily squeezed than if the sub-pads were larger.

11.3   TARANTULAS AND SPIDER-MAN AND SPIDER WEBS

Apparently, tarantulas don't like falling from a great height. They take care not to slip off whatever they are climbing. You will not be surprised to know that their primary climbing mechanism is via gecko-style adhesion. It seems, however, that for hanging around for a long time, the gecko mechanism is insufficient; they have added hooks to their legs for safe mechanical attachment. The hooks and gecko adhesion are still not enough for the security the tarantulas crave so they have one more mechanism. Yes, they are like Spider-Man and can shoot out threads to stop themselves from falling.

Spiders in general have developed a number of adhesive strategies for trapping their prey. Some put blobs of adhesive at the end of a dangling thread. There is enough liquid adhesive in the blob to properly engulf enough parts of the prey to prevent escape. Other spiders make some of the threads in their web sticky enough to trap the prey, while using enough non-sticky threads to allow the spider to move around without getting trapped in its own web. Because the amount of adhesive is necessarily limited, larger prey such as moths can escape, leaving a few ripped-off scales. Not surprisingly, some spiders combine blobs and webs in complex structures to extend the range of prey they can capture.

Many types of spider silk have a nano-fibre surface that can provide basic gecko-style adhesion (i.e. it's just van der Waals attraction aided by being able to create lots of surface contact) while also providing physical entanglement – nano-filament structures on an insect's legs becoming entangled in the spider's nano-fibres.

Previously, we have seen that some natural-based adhesives are proteins such a collagen or starches such as flour paste. Spiders seem to have chosen a mix of the two, glycoproteins, where the "glyco-" part means lots of sugar molecules stuck together in a polymer, just like starch. The spider has two problems with this glue, one of which it can readily solve and the other of which it cannot.

- The adhesive is based on water and needs to remain sticky for hours or days, so the water has to be encouraged to stay.

The adhesive therefore contains lots of small, water-loving (hygroscopic) molecules that are easily available as the by-product of producing molecules such as proteins. They are remarkably similar to the "natural moisturizing factors" that keep the surface of the human skin hydrated.

- Whether it likes it or not, a layer of liquid adhesive covering a thin fibre cannot remain as a thin layer, it naturally flows to form a string of beads along the fibre. This is due to the Rayleigh instability and is the cause of those beautiful early misty morning glistening spider webs. The water that has condensed onto the surface also breaks up into regularly-spaced beads that sparkle in the early morning sunshine. Because the spider cannot fight the physics of the Rayleigh instability, it no doubt has worked out how to take advantage of these sticky beads to help better trap the insects.

There is one more trick to spider webs that is related to the key idea of dissipation. Imagine first a totally rigid web. An insect flying into it might very well bounce back instantly, with no chance of sticking. Now imagine a totally elastic web. The insect would stretch the web when it flew into it then be flung out of the web when the elastic snapped back. Instead the webs are designed to be dissipative – they move under the impact, absorbing the energy so there is no spring-back effect to eject the prey.

11.4   CLIMBING ROBOTS

During the couple of decades of intensive work on the physics of gecko-style locomotion, there was always the expectation that the work would translate nicely into climbing robots. The general ideal was something the size and weight of a gecko, equipped with a camera, zipping around buildings or disaster zones, coping with heights and different surfaces, with the ability to get into tight spaces and look for further dangers or trapped people.

While the physics is clear, it has not yet proved possible to find ways to get the compliance (legs, toes, lamellae, setae, spatulas) from mm to nm scales *and* to have the fluent motions of the geckos *and* to have a power source capable of sustaining the robot for a significant time without a tether.

Yes, it is possible to use some gecko-style adhesion that allows a robot to do a difficult, one-off climb of a glass-clad building, but that is a long way from using the general-purpose capabilities inherent to a real gecko's adhesion system.

There is an interesting alternative, even if it dispenses with all the physics and is disturbing to some of us: hack the brain of a real gecko by sending it signals so it "chooses" to go where we want it to go and do what we want it to do.

11.5   UNDERWATER ADHESION

The most researched underwater adhesive system is that used by mussels to stick to rocks. We have all the necessary ideas to understand how mussel adhesion works and can go through them step by step.

- Rocks provide uncertain anchors of varying shapes, chemistries, (un)cleanliness and chemistry. The first imperative of the mussel is not to rely too heavily on one spot. The answer is to put out lots of individual threads, the ends of which stick more or less well to the rocks and which can cope with stresses coming in many directions as the currents and waves attack the adhesion points.
- Perfect surface contact is a necessity for any robust adhesive, so the mussel extrudes a liquid adhesive which makes perfect contact before it sets. Doing this in sloshing sea water would be rather a challenge. Contrary to naïve myths about amazing mussel adhesives, the mussel first builds a microdam around the bit of rock and fills it with its own, controlled, aqueous system into which the various chemicals are extruded and where a set of complex processes yield an adhesive in contact and ready to react.
- Surface energy cannot possibly survive the ocean's attacks on the adhesive, therefore strong adhesion is required via chemical bonding to the rock. The mussel knows that too much of a good thing is a bad thing and creates only a modest number of bonds. It also knows that dissipation is key, so it connects the bonds via a crosslinked protein network. Any chemist could suggest three systems that might do the job.

○ We have already encountered two systems that might achieve this – the silanes and the carboxylic acids. A smart chemist might choose a "chelating" system (named after crab claws) using two carboxylic acids that can each grab onto the rock. We saw an example of this (maleic acid polymers) in Chapter 5.

○ However, the mussels have chosen a chemistry that chelates and can also crosslink. This is the chemistry of catechols. These are a type of phenol (as used in disinfectants) but with two –OH groups next to each other. These two groups can chelate onto the rocks. They also happen to be readily convertible to reactive quinones which can then be involved in the necessary crosslinking reactions to give the required dissipative network.

The mussel has chosen the right science. It uses multiple attachments for compliance and resilience, it gets good surface energy contact via a liquid glue that subsequently sets, it provides sufficient chemical bonds for entanglement, chosen for their built-in safety feature (chelation) and uses a modestly crosslinked network to provide the dissipation (Figure 11.2).

It is interesting to ask how much inspiration we should take from mussel adhesion. The usual stance is that we should try to mimic it because it is natural and natural is automatically good. I disagree. Let us compare the natural and artificial systems.

• It is often said that it is amazing that the mussel can apply the adhesive under water. In fact, it applies it in an

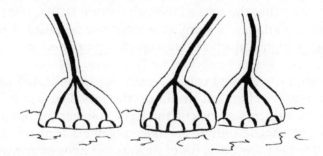

Figure 11.2 The mussel uses multiple fibres, ending in multiple sub-fibres that stick to the rock via entangled chemical bonds.

environment that it carefully creates via a hugely complex process. We have many ways that we can achieve the same result with far less expenditure of time, chemicals and energy. Silane-terminated adhesives work perfectly well under water.

- Mussel adhesive is built from peptides with many different side-groups and functionalities that provide nothing useful. A chemical alternative would use the minimum amount of the simplest-possible monomers, making it resource efficient. In this sense, biology is wasteful of resources.

- Although the mussel adhesive is assembled at ambient temperatures in an aqueous environment, this can only be done by a vast array of energy-consuming processes that slowly convert the energy in mussel food into chemistry, spewing out waste from each step. We don't notice this because the process is on a miniscule scale. We notice the heat and waste when we make a ton of artificial adhesive because it is concentrated in space and time. Modern industry is very good at not wasting the energy required for the synthesis (it is expensive) and in recycling things like solvents (which are also expensive).

- Mussel adhesive uses harmful chemicals. In general, catechols (like phenols) are nasty and corrosive. The known effects on humans of the main catechol used by the mussel include: hypotension; arrhythmias; nausea; gastro-intestinal bleeding; hair loss; hallucinations; narcolepsy. That catechol is DOPA, dihydroxyphenylalanine. You might be familiar with DOPA as it is a potent brain chemical used, for example, in the treatment of Parkinson's disease. It sounds very nice that the mussel uses all-natural DOPA, but I doubt that any adhesives supplier would ever be allowed to have such a high concentration of such a potent chemical.

Why does the mussel use, for example, complicated proteins when a simpler chemical would do the job? Because it has no choice. We can choose exactly the right polymer for the job; the mussel can choose only from those polymers that can be synthesized from a few basic units such as peptides or sugars. This is a strength and weakness of nature. We can marvel, for example, at the strength of many biological materials such as spider silk, yet there is a lot to be said for building bridges from steel and houses from bricks and windows from glass, materials forever inaccessible to nature's amazing but limited repertoire of materials.

Don't get me wrong. I am in awe of what natural systems can achieve and for many years I have been part of the bio-mimetic community that seeks inspiration from biology's solutions to difficult problems. I have made gecko tapes, moth-eye antireflection structures, shark skin easy-flow surfaces, Lotus Effect self-cleaning surfaces, photonic crystals (the trick used by nature to create iridescent colours). In each case, the science has been deeply fascinating and the practical outcomes deeply disappointing. A different way to seek bio-inspired solutions to problems is encapsulated in a technique called TRIZ (this is a Russian acronym that can be translated as "theory of the resolution of invention-related tasks"). TRIZ has codified the many imaginative ways that biology solves problems (plus imaginative ways developed by smart inventors) and invites engineers to solve problems by thinking differently. The solutions they come up with are still artificial (they might continue to use steel, for example), but they solve the problem by using elegant, bio-inspired principles as alternatives to the obvious brute-force solution.

11.6 FIGHTING ADHESION

I admit it in public. I have engaged in chemical warfare with a cunning and resourceful enemy. Just because something is natural does not mean that it is nice. My enemy was 100% natural and thoroughly nasty. I tried to be nice, but in the end it was them or me and it wasn't going to be me.

"Biofilm" sounds a harmless term, maybe even friendly. It is a thin film made up of an inter-dependent array of natural organisms such as bacteria or fungi, surrounded by a polymeric protective film to keep the community safe and secure. In my specific case of chemical warfare, the biofilm was a green mould growing in a less than ideally ventilated bathroom. How hard can it be to remove a few cells from the surface of a bath, wall or ceiling? It turns out to be very hard. I zapped them with fairly strong bleach which chemically attacks just about everything in a biofilm, destroying the whole system, producing detritus that could be easily removed. I then improved the ventilation and made sure the fresh paint contained a mould-resistant agent, just in case – more chemical warfare.

In my case the biofilm was an unsightly green stain. For those with, for example, dental plaque, gingivitis, urinary tract infections, cystic fibrosis or infected heart valves, the biofilms are somewhere along the spectrum from a nuisance, to painful, to lethal.

As we know, adhesion is a property of the system and biofilm is an amazing system. It is a community of organisms, where even a single organism might manifest itself as different sub-populations with different sub-specialities (Figure 11.3).

- Some parts of the community provide the intimate surface contact that is the minimum requirement for adhesion. This is not as easy as it sounds. You can readily prevent the build-up of some *specific* organisms onto a surface by providing a shape the organism happens not to like. Yes, some bacteria don't like pointy features or cubic features or cylindrical features. The way they can sense the shape of a surface is discussed in the next section. Yet there is always some member of a biofilm community that is happy with any given surface feature, so the community quickly covers the surface with perfect contact. I was once involved in a project to reduce bio-fouling for structures in the sea. An approach based purely on surface structures would have been wonderful compared to one based on chemicals. But, whilst the

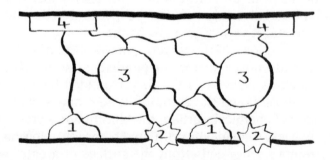

Figure 11.3   Biofilm is a community of different organisms (and sub-organisms) that each play a role throughout the film.

surfaces could resist a large array of marine organisms, there would always be one that could cope with it, and once there was one, the others could join in.

- Then the community can unleash whatever chemistries give intermingling (into soft surfaces), entanglement (into even softer, more open surfaces) or chemical bonding onto a surface linked via the biofilm's own polymers into a dissipative network.

- Finally, the community has to stick itself together with a gunky hydrogel made from proteins, polysaccharides, lipids and DNA – whatever comes to hand and provides the balance of hardness and resilience required to meet whatever environmental threats happen to be around – such as my scrubbing brush and the bathroom cleaner chemicals I had first tried to use. The community doesn't "know" in advance what it wants. If it made a wrong guess and my scrubbing brush and cleaner destroyed a large part of the community, a little experimental community that happens to survive quickly replicates itself. My sense of victory that a bit of muscle and surfactant power was all it took is disproved a few days later when I see that the "same" film has bounced back. When I try the same brush and cleaner, the results are disappointing, and lead to another round of response from the biofilm that makes the film even more resistant.

The community is equally resistant to biological warfare. An antibiotic might zap 99% of one part of the community. The 1% that were resistant bounce back to some extent, while other parts of the community take over some of that space. After just a few cycles of biological warfare the battle is lost, and the community has total resistance. Why is this? Because there is biological warfare going on within the biofilm. Each organism has to fight against being overwhelmed by a more vigorous one. There is a constant battle of natural antibiotics being developed, overcome, modified and so forth. It isn't we humans who invented antibiotics, we found them in nature where they are a part of a balance of survival and cooperation. And it isn't that the organisms have evolved to fight *our* antibiotics – they have evolved to fight *all* antibiotics, as best they can.

A specific example of this dynamic action is how dental plaque develops into and is locked in as tartar – the solid gunk that has to be scraped from the teeth. A softer community is presumably not so capable of withstanding the tough environment of the mouth. By using the plentiful supply of calcium phosphate salts in the saliva, a hardened environment can be created. The process is very much like the creation of underwater cement – the phosphates can crystallize into a solid network. To gain strength the network needs fillers equivalent to the sand and gravel of real cement. Biology always uses whatever is available and in this case that means dead cell debris. Just as dentine is made tough by having flexible collagen/proteins between the hydroxyapatite particles, so these ex-cellular fillers toughen the tartar. The biofilm community doesn't "know" which cells to kill off and which cells will promote the phosphate cement formation. Those parts of the community that happen to create such a microenvironment can flourish while other parts get swept away with a toothbrush. The trick is for the community to kill off enough of its own members to provide the cement fillers while keeping the rest of the community in good shape to continue building, by being securely attached to their own concrete via some complex protein/starch network.

There are three standard ways to fight with the adhesion of biofilms:

1. Don't provide the conditions for them to form in the first place. I should have improved the bathroom ventilation in the old house *before* problems appeared.
2. Zap them in their very early stages. I should have noticed a slight green film and dealt with it early on.
3. Total chemical, biological and mechanical warfare. What I had to do to solve the problem.

A fourth way is much like the newer approaches to gut problems: try to encourage the ecosystem to work *with* you rather than *against* you, encourage good bacteria against bad bacteria. For example, although dental plaque is unsightly, a plaque that didn't attack your teeth with acid would be a good defence against an alternative plaque that spewed out lots of acid. It is not easy to flip a biofilm community from a "bad" state into a "good" state. The naïve idea that dumping a large excess of one type of bacteria will help flip the community rarely works. The current community merely needs to excrete one type of biological defence to overwhelm these new invaders. Finding a set of bacteria that, together, can transform the community is harder in the short term and more likely to succeed in the long term.

11.7   HOLDING OURSELVES TOGETHER

What is it that keeps our bodies together? Our first thoughts tend to be along the lines of "there's a skeleton for the rigid outline and muscles and tendons to attach the soft bits to the hard bits". Even if we accepted that skeletons, muscles and tendons pose no "adhesion" issues, we still have the question of why, say, our liver stays together as an identifiable structure.

It turns out that most cells in our bodies are "adherent" cells in that not only do they like to adhere, but they die if they do not have something to adhere to. The death of non-adhering cells is a good thing, otherwise rogue cells would forever be wandering off and sticking in the wrong place. There is another, "non-adherent" cell type, of which the red blood cell is a key example. Those who are unfortunate enough to have somewhat sticky red blood cells need careful doses of warfarin to keep their blood flowing. There is a third type of cell which is switchable. Our white blood cells must wander happily through our blood system then switch to adhering mode when it is time to fix a problem at a specific location.

Like most people, I had never given the question of cell adhesion any thought until I was invited to a conference where this was a key topic. Conferences are made up of two components. The formal talks are where you sit quietly for 30 minutes and usually learn not very much. The informal talks are those held over coffee and meals and these are where you learn the most.

In terms of "why are we held together" I quickly learned over coffee that there were two irreconcilable camps. In one set of conversations it was irrefutably the case that we were held together by basic adhesion physics, end of argument. In another set it was clear that without the fact that adhesion molecule X had a controlled hydrogen bond with cell molecule Y, our bodies would fall apart. One camp was all about general principles, the other was all about specifics.

I will attempt to present both sides of the argument, starting with the general principles approach to which we can give a simple name.

11.8   THE KENDALL FORMULA

Prof. Kendall's book "Adhesion of Cells, Viruses and Nano-particles" wastes no time in telling us the answer. In the Preface we learn that cells adhere via the Kendall formula, $F = K \times [W \times E \times d^3]^{\frac{1}{2}}$, where F is the force, K is a constant, W is a work of adhesion, E is a modulus and d is a relevant dimension. Because adhesion is a property of the system, the basic surface energy work of adhesion (W) is amplified by the modulus and dimensional effects. Although he acknowledges that W depends on adhesion molecules, the effects of E and d are generally far larger. This takes us back to Chapter 4 where peel, lap shear and butt joint forces for a simple surface energy adhesion varied by many orders of magnitude because modulus and dimensional effects are so significant for lap and butt modes. Indeed, the formulae used there were also developed by Prof. Kendall.

The Kendall approach is to keep things as simple as possible. A large chunk of the cell adhesion book relies on experiments with rubber and glass, showing that many of the essential features of cell adhesion can be identified without the need for complex biological arguments. Water, proteins and surfactants are generally identified as contributing to a *reduction* of adhesion, except for gecko adhesion where a bit of water probably makes the keratin in the gecko pad a little more conformal and therefore able to generate more adhesion via perfect contact.

The effects that we have frequently encountered in this book, such as dissipation on the nm or µm scale, appear later in the Kendall book, showing 100× increases in adhesion. There are also delightful demonstrations showing that the smooth peel of a piece of rubber almost comes to a stop when the rubber suddenly increases in thickness, even though the peel force of that thicker rubber is, of course, identical. The reason is that for a moment the peel force is diverted into bending the thicker rubber, temporarily delaying the peel. By analogy, a biological system could provide crack-stopping ability simply by having local changes in thickness. Crack stopping via breaks in the adhesive contact are also demonstrated. Nature is happy to use simple physics precisely because it can make large practical gains for small effort, such as making a part a bit thicker.

So far, cellular adhesion is nothing special, merely confirming what we already know. A slight twist to the idea that adhesion is a property of the system is the fact that how cells adhere to a surface often depends less on the chemical nature of the surface and more on the modulus of the surface. For example, tests that compare, for scientific convenience, hydrophobic polystyrene with hydrophilic glass show no significant difference in cellular adhesion. As far as the cells are concerned the surfaces are the same because their moduli are indistinguishable, in the GPa range. On the other hand, cell cultures sense the difference between a GPa surface and a kPa surface and grow very differently on a soft surface. And, as mentioned near the start of this section, cells floating free of a surface cannot survive – cells self-destruct if they don't feel adhesion forces.

How do cells "know" or feel that they are sticking to something? They contain actin/myosin fibres, which are basically fibres that can contract just like a normal muscle. A network of fibres within the cell can therefore feel the force of their combined attachments to the external surface. A stiff surface will resist the fibres' contraction forces very differently from a soft one, while there is no resistance if the cell is floating free, so the cell knows to self-destruct.

11.9 IT'S ALL ABOUT SPECIFIC MOLECULES

The other side of the story – the specificity – starts at this point. The muscle fibres within the cells prefer to connect to surfaces via specific adhesion molecules with names like integrins, selectins or cadherins. Without such specificity, cellular adhesion would be a mess. So, the argument goes, these molecules have special adhesion properties. This is undoubtedly true, but the "special" part is not some amazing adhesion mechanism such as the much-loved lock-and-key idea whereby the specific interactions provide an otherwise unattainable adhesion force.

To understand why these molecules are special we need to identify the two competing problems faced by biology:

- General van der Waals adhesion via perfect surface contact could be devastating because it is probably more than enough to get everything to stick together indiscriminately. Indeed, some biological systems go to great lengths to create spikey surfaces that avoid perfect surface contact.
- The opposite problem is that in an aqueous environment full of random large molecules, the data show that getting *anything* to stick is very hard. If you try to bring two surfaces close enough together (~1 nm) to get good surface adhesion, you have to try hard to push away water molecules and try even harder to push away large biological molecules. And in any case, the ordinary biological surfaces sticking out into water are negatively charged and repel each other (DLVO). We saw this effect in Chapter 10 where this anti-adhesion effect is necessary for hip joints. Turning this self-repulsion trick around, a generic and effective class of biocides are "quaternary amine" polymers which are positively charged. They head straight to the negatively charged surface and cause havoc by associating in ways the biological system does not expect.

To avoid sticking too hard, which would render the cells too immobile to function well, the trick is to provide just enough sites that stick with general vdW strengths. The "just enough" involves adhesion molecules like the integrins forming modest regions of self-association around the end of, say, the muscle fibres, producing multiple, large-enough attachment sites. To get "general vdW strengths" these sites must be able to approach the complementary site, efficiently sweeping away interfering water molecules and competing biopolymers, and possessing enough positive/negative complementary interactions to overcome the DLVO barrier. In this way the molecules are smart not because of strong adhesion but because they manage to achieve their controlled, limited adhesion amidst the chaos of an aqueous biological system (Figure 11.4).

Once we have this attachment, the rest of the system must provide the dissipation needed for robust adhesion. Bonding via

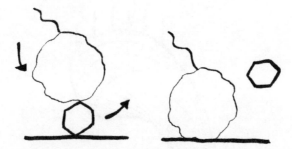

Figure 11.4 The large integrin blob is attached to a cell fibre and can attach to the surface below. There's nothing special about its attachment. What it is good at is kicking out stuff that is already at the surface.

a network of muscle fibres is a variant on the entanglement theme. For general crosslinked entanglement in polymers, the length scales are nm, for PSAs the length scales are μm and mm; the entanglements via muscle fibres are in the tens of nm domain. The important factor is not so much the domain size but the existence of a robust dissipation mechanism.

11.10   ADAPTIVE ADHESION

There is a wonderful book from 1917 by D'Arcy Wentworth Thompson called *On Growth and Form*. Thompson was rather tired of biological explanations that were, well, biological. The point of the book was that the laws of physics are everywhere and that many biological structures exist via straightforward physics rather than anything especially biological. I remember reading about the cell structures inside a bone. They looked as though cells had been placed at exactly the right place to be able to absorb the specific stresses of that specific part of that specific bone. A modern engineer with computer aided design (a luxury Wentworth Thompson didn't have) would place load-bearing elements in exactly the same spots. How did the bone perform the calculations and then know where to put the cells? It didn't. The bone just responds to whatever is happening to it over the decades. The cells are able to detect stresses (via the fibres necessary for adhesion control), and wherever there are large stresses, more cells grow, to reduce those stresses. If there are low stresses, enough cells will die off until the stresses are back to normal (Figure 11.5).

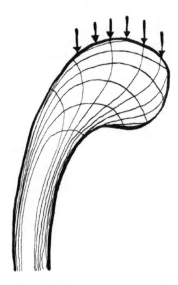

Figure 11.5 This bone-like structure has a large load along the top. It automatically creates the strongest cellular structures at the points where the stresses are largest.

This automatic response to stresses applies across many biological structures such as tree branches, in each case responding to stresses and placing resources where they are needed most. The principle applies to us throughout our lives. Those who don't exercise enough end up with weaker bones because the body doesn't bother to use up precious resources where they are not needed. Conversely, modest exercise over time can build up the right bones and muscles and make our bodies more resilient to the accidents of life.

Any form of strong adhesion in the responsive parts of trees or humans would be a disaster. Sure, organisms can create classical strong adhesives or ceramics in parts that do not have to respond. In general, however, it is the weak, dynamic form of adhesion that is needed to keep an organism able to function in an ever-changing environment.

Those adhesion molecules such as integrins are *not* providing any sort of strong adhesion. Instead they are providing *smart* adhesion – just the amount needed to cope with today's challenges with a margin of error sufficient to withstand a storm (for a tree) or an unexpected run for a bus (for human bones).

I started this chapter being rather dismissive of attempts to create bio-inspired adhesives. The more you look into bio-adhesives, the more you realize how amazing, how smart they are. My problem with bio-inspired adhesives does not lie in the adhesives themselves. My problem is that in general we want adhesives to solve different types of problems. We do not want to be providing a continual supply of materials and energy to keep an adaptive adhesive system functional, whilst removing waste as it adapts. We want to define our problem, do the engineering, chemistry and physics, and provide a one-off solution that needs no further attention. As I said early on in the book, I do *not* want the adhesive that sticks the wing on my aircraft to be doing the stress calculations for me as we fly along. I want an engineer to have done the calculations, taking into account a margin of error.

The problem fundamentally affecting those working on bio-inspired adhesives is to identify a need for which the classical adhesive approach cannot work. Engineered adhesives are capable of doing many things that biology cannot do, and their success makes it unlikely that most bio-adhesive projects will provide mainstream products. However, sometimes something like a weak, adaptive adhesive system will be superior. This is likely to be the case for some specialist adhesives used in medicine because the infrastructure to provide the necessary resources for an adaptive adhesive already exist. In the 3D adhesives chapter we saw that a crudely-constructed prototype organ such as a kidney must be able to adapt as the wrong cells die off and new ones take their place. For non-biological applications it is much harder to identify the opportunities.

An important factor often excluded from discussions about real bio-adhesive systems is that their shortcomings can be fixed by an active repair system, something not possible with most bio-inspired adhesives. A mussel can replace an attachment link if one is lost in a storm. This requires a diversion of energy and nutrients from one process (growing, for example) to another (fixing the adhesive), which is the sort of diversion that natural systems perform all the time.

Those who have been inspired by such examples to create "self-healing systems" generally adopt a method necessarily more simplistic than that used by nature. These usually work by having a pool of liquid (or semi-solid) within the system which can flow and react into, say, a crack and heal the system. The number of such systems that have become practical reality is small because of the two obvious problems:

1. The pools of liquid available for healing are a waste of strength and resource for most of their existence
2. A repeated crack in the same place (because, for example, the stresses that caused the initial failure are still there) cannot be fixed because the pool has been used up

I am not saying that smart, self-healing, self-strengthening systems inspired by nature are impossible or undesirable. I *am* saying that obviously bad ideas wrapped up in a claim of being bio-inspired are still bad ideas.

Now that we have reached the end of our journey through adhesion science, I want to finish with a reflection on the most important principle of all – the unity of science.

11.11   THE UNITY OF (ADHESION) SCIENCE

My day job often features people throwing difficult scientific problems at me in areas where I have no direct expertise. What I find astonishing is that after the first moments of panic, some general scientific principle from another area suggests itself as being relevant. I give that a try, and although it may not help in itself, it stimulates discussions which lead, eventually, to a helpful response. Further discussions might result in clear evidence that the first idea is *not* relevant – and a clear negative in science is of great value because there is no need to waste time discussing it further. This frees the mind to look for another principle. After a few cycles of ideas, it generally works out that the combination of my outsider's knowledge and their insiders' expertise is enough to find a way forward.

This is only possible because of the unity of science. Something I know from solubility can be applied to adhesion. Something I know from coatings can be applied to surfactants. At one time, the vague memory of an obscure undergraduate lecture was enough to give me the idea that was the basis of my PhD work. Once, when, by chance (I was at that institute for unrelated reasons) I found myself standing just 50 cm away from a seriously deadly cobra being held by a snake expert, I unexpectedly found myself pointing out some fluid mechanics I happened to know from a very different context. This clearly refuted the expert's ideas of how the snake ejected its venom and allowed a fresh set of ideas to emerge.

This whole book is based on a handful of rather straightforward principles. The focus on each of the principles shifts as we look at different adhesion problems, but we need to keep juggling them all. It might be right to focus on strength, but a mistake to forget about toughness. Dissipation might be the key, but flow that is too easy will be undesirable. We might be concerned with temperature, but we should not forget about time. We might want more chemical bonds, but we don't want to create a brittle glass. Our adhesive might have the perfect modulus, but that of our adherend might be the problem.

The unity of adhesion science resides in the core principle that adhesion is a property of the system. I like including a guarantee in my books, and for this one it comes right at the end. I guarantee that if you think as much about the system as you do about the adhesive, then you will have greatly increased your chances of your adherends sticking together.

Subject Index